Survival Analysis Using S

Analysis of Time-to-Event Data

CHAPMAN & HALL/CRC
Texts in Statistical Science Series

Series Editors
Chris Chatfield, *University of Bath, UK*
Martin Tanner, *Northwestern University, USA*
Jim Zidek, *University of British Columbia, Canada*

Survival Analysis Using S

Analysis of Time-to-Event Data

Mara Tableman
Jong Sung Kim

with a contribution from
Stephen Portnoy

CHAPMAN & HALL/CRC

A CRC Press Company
Boca Raton   London   New York   Washington, D.C.

Library of Congress Cataloging-in-Publication Data

Tableman, Mara.
  Survival analysis using S : analysis of time-to-event data / Mara Tableman & Jong Sung
Kim with a contribution from Stephen Portnoy.
     p. cm. — (Texts in statistical science)
  Includes bibliographical references and index.
  ISBN 1-58488-408-8 (alk. paper)
   1. Failure time data analysis—Data processing. 2. Survival analysis (Biometry)—Data
processing. 3. S (Computer system) I. Kim, Jong Sung. II. Title. III. Series.

  QA276.4.T32 2003
  519.5—dc21   2003053061

Visit the CRC Press Web site at www.crcpress.com

© 2004 by Chapman & Hall/CRC

No claim to original U.S. Government works
International Standard Book Number 1-58488-408-8
Library of Congress Card Number 2003053061
Printed in the United States of America 1 2 3 4 5 6 7 8 9 0
Printed on acid-free paper

To Werner Stahel, my longtime friend and colleague
To my wife Seokju, son Thomas, and daughter Jennifer
To my wife Esther

Contents

Preface

The earliest version of this book was a set of lecture notes for a block course in Survival Analysis in the program, Nachdiplomkurs in angewandter Statistik (post graduate course in applied statistics), at the Swiss Federal Institute of Technology (ETH), Zürich, Summer 1999. This particular block is allotted five half-day sessions delivered on three consecutive Mondays. Each half-day session consists of two 45 minutes lectures followed by an hour and half computer lab where first S, but now R is the required statistical computing language.

This course's diverse audience has minimally two attributes in common, a university diploma (equivalent to Master's degree) and a desire to learn data analytic tools and statistical methodologies to be applied in their work places and in research areas such as plant science, biological science, engineering and computer sciences, medical science, epidemiology and other health sciences, statistical science, and applied mathematics.

This book is written with two main goals in mind: to serve the diverse audience described above and to teach how to use and program in S/R to carry out survival analyses. However, the pedagogic style can serve as a self-learning text and lends itself to be used for a WebCT course as S/R code and output are woven into the text. Throughout the book, we emphasize exploratory data analytic tools. The S functions written by the authors and the data sets analyzed are available for download at the publisher's website, www.crcpress.com. The R language is the "free" S and is available to the public for download at www.r-project.org.

This book attempts to introduce the field of survival analysis in a coherent manner which captures the spirit of the methods without getting too embroiled in theoretical technicalities. Hence, the minimum prerequisites are a standard pre-calculus first course in probability and statistics and a course in applied linear regression models. No background in S/R is required.

The book offers two types of exercises: A and B. Type A, titled *Applications*, is designed for students with minimum prerequisites in mind. Type B, titled *Theory and WHY!*, is designed for graduate students with a first course in mathematical statistics; for example, Hogg and Craig (1995). Students who had a course in applied probability models, for example Ross (2000), have an opportunity to apply knowledge of stochastic integrals and Gaussian processes.

The WHY! is our trademark and we use it throughout. These exercises also help students to practice their skills in mathematical proofs.

For an overview of chapter by chapter material, read our table of contents. We purposely designed it to highlight the key methods and features contained in each chapter. We devote a good portion of the text to resurrect the importance of using parametric models, as they can be very useful for prediction and in small sample settings. The newer, very flexible nonparametric procedures along with model building and data diagnostics are carefully detailed. Furthermore, we discuss cut point analysis so often desired by medical researchers. We introduce *bootstrap validation of cut point analysis*, which shows its robustness. This also provides a useful guide to decide whether or not to discretize a continuous prognostic factor. We compare a Cox PH model and an extended Cox model and provide an S program that incorporates time-dependent covariates based on a counting process formulation. Competing risks and the necessity of the cumulative incidence estimator are succinctly discussed. Although our emphasis is analyzing right-censored data, we provide an analysis of left-truncated and right-censored data with researchers in the public health arena in mind. Finally, a nonparametric regression approach is introduced for the first time. This methodology is called *Censored Regression Quantiles* and is developed to identify important forms of population heterogeneity and to detect departures from traditional Cox models. Developed by Stephen Portnoy (2003), it provides a valuable complement to traditional Cox proportional hazards approaches by generalizing the Kaplan-Meier estimator (Chapter 2) to regression models for conditional quantiles. That is, the idea of univariate quantiles for right-censored data is extended to the linear regression setting in Chapter 8.

The level of this book is pitched between Kleinbaum (1995) and Klein and Moeschberger (1997). Its level is comparable to Hosmer and Lemeshow (1999). Readers of our text will find that exercises in Kleinbaum can be easily handled by our S/R examples. Although Klein and Moeschberger target a more mathematically sophisticated audience than our book does, a reader can transition with comfort from our book to theirs as these two books have much in common. Many of our S/R examples can be directly applied to solve problems in Klein and Moeschberger. We recommend the first six chapters for universities with a ten-week quarter system, and seven chapters for universities on an approximate fifteen-week semester system. Chapters 7 and 8 can serve as supplemental material for a faster paced course or as special topics in more advanced courses.

We were influenced by Kleinbaum's pedagogic style, which includes objectives and a summary of results for each analysis. We are students of earlier books written by Kalbfleisch and Prentice (1980), Miller (1981), Lawless (1982), Cox and Oakes (1984), Kleinbaum (1995), and Klein and Moeschberger (1997).

The authors wish to thank the graduate students at the ETH, Portland State

University, and Oregon Health & Science University, whose feedback greatly improved the presentation of the material. The authors greatly appreciate Stephen Portnoy for contributing not only his beautiful Chapter 8 but also his wisdom, good humor, kindness, enthusiasm for this project, and friendship. The authors together with Stephen Portnoy extend their gratitude to Roger Koenker for his careful and critical reading of the material in Chapter 8. Stephen Portnoy also wishes to acknowledge his guidance and inspiration over the past quarter century. Further, the authors wish to thank the Series Editor Jim Zidek for the opportunity to join the family of authors in the Chapman & Hall/CRC statistical science series. We thank Kirsty Stroud, our editor, and her staff for their professional support throughout the entire process. Finally, we thank the external reviewers whose in-depth critique of a crude version provided a path for us to follow.

The author, Jong Sung Kim, wishes to thank Dongseok Choi, Marc Feldesman, and Dan Streeter for their help with a number of computational issues. I also extend thanks to Mara Tableman for having invited me to be involved in this project. This opportunity has enriched my academic and personal life. I thank my advisor, Professor Jian Huang, for his continued support. Most importantly, without boundless support from my wife Seokju, my son Thomas, and my daughter Jennifer, I could not have made this achievement so early in my academic career.

The author, Mara Tableman, wishes to thank Michael Lasarev for assistance with some computational issues and Brad Crain for assuming one course from my two-course teaching load. This enabled me to meet the publisher's preferred deadline for the completed manuscript. I wish to thank my young, very talented colleague, Jong Sung Kim, for joining this project as a coauthor. Our discussions and his insight enhanced the material and quality of this manuscript. His passion for statistics and life brought much to this project. I also extend my deepest gratitude to the Seminar für Statistik, ETH, for inviting me to be a lecturer in the Nachdiplomkurs in angewandter Statistik and for having provided me a nurturing home for my sabbatical year and a summertime residence. I'm blessed to have their continued support and interest in my academic career. Finally, as always, I so appreciate the continued support of my "Doktorvater" Tom Hettmansperger, whose wisdom and kindness are a continued source of inspiration.

CHAPTER 1

Introduction

The primary purpose of a survival analysis is to **model and analyze time-to-event data**; that is, data that have as a principal endpoint the time when an event occurs. Such events are generally referred to as *"failures."* Some examples are time until an electrical component fails, time to first recurrence of a tumor (i.e., length of remission) after initial treatment, time to death, time to the learning of a skill, and promotion times for employees.

In these examples we can see that it is possible that a *"failure"* time will not be observed either by deliberate design or due to **random censoring**. This occurs, for example, if a patient is still alive at the end of a clinical trial period or has moved away. The necessity of obtaining methods of analysis that accommodate censoring is the primary reason for developing specialized models and procedures for failure time data. **Survival analysis is the modern name given to the collection of statistical procedures which accommodate time-to-event censored data.** Prior to these new procedures, incomplete data were treated as missing data and omitted from the analysis. This resulted in the loss of the partial information obtained and in introducing serious systematic error (bias) in estimated quantities. This, of course, lowers the efficacy of the study. The procedures discussed here avoid bias and are more powerful as they utilize the partial information available on a subject or item.

This book attempts to introduce the field of survival analysis in a coherent manner which captures the spirit of the methods without getting too embroiled in the theoretical technicalities. Presented here are some frequently used parametric models and methods; and the newer, very fashionable, due to their flexibility and power, nonparametric procedures. The statistical tools treated are applicable to data from medical clinical trials, public health, epidemiology, engineering, economics, psychology, and demography as well.

Objectives of this chapter:

After studying Chapter 1, the student should be able to:

1. Recognize and describe the type of problem addressed by a survival analysis.
2. Define, recognize, and interpret a **survivor function**.

1

3. Define, recognize, and interpret a **hazard function**.

4. Describe the relationship between a survivor function and hazard function.

5. Interpret or compare examples of survivor or hazard curves.

6. Define what is meant by **censored data**.

7. Define or recognize six **censoring models** and two **truncation models**.

8. Derive the likelihood functions of these models.

9. Give three reasons why data may be randomly censored.

10. State the **three goals of a survival analysis**.

1.1 Motivation - two examples

Example 1. AML study

The data presented in Table 1.1 are preliminary results from a clinical trial to evaluate the efficacy of maintenance chemotherapy for acute myelogenous leukemia (AML). The study was conducted by Embury *et al.* (1977) at Stanford University. After reaching a status of remission through treatment by chemotherapy, the patients who entered the study were assigned randomly to two groups. The first group received maintenance chemotherapy; the second, or control, group did not. The objective of the trial was to see if maintenance chemotherapy prolonged the time until relapse.

Table 1.1: *Data for the AML maintenance study. A + indicates a censored value*

Group	Length of complete remission (in weeks)
Maintained	9, 13, 13+, 18, 23, 28+, 31, 34, 45+, 48, 161+
Nonmaintained	5, 5, 8, 8, 12, 16+, 23, 27, 30, 33, 43, 45

Example 2. CNS lymphoma data

The data result from an observational clinical study conducted at Oregon Health Sciences University (OHSU). The findings from this study are summarized in Dahlborg *et al.* (1996). Fifty-eight non-AIDS patients with central nervous system (CNS) lymphoma were treated at OHSU from January 1982 through March of 1992. Group 1 patients (n=19) received cranial radiation prior to referral for blood-brain barrier disruption (BBBD) chemotherapy treatment; Group 0 (n=39) received, as their initial treatment, the BBBD chemotherapy treatment. Radiographic tumor response and survival were evaluated. Table 1.2 describes the variables obtained for each patient.

Of primary interest was to compare survival time between the two groups. On

Table 1.2:   *The variables in the CNS lymphoma example*

1. PT.NUMBER: patient number
2. Group: 1=prior radiation; 0=no prior radiation with respect to 1st blood brain-barrier disruption(BBBD) procedure to
3. Sex: 1=female; 0=male
4. Age: at time of 1st BBBD, recorded in years
5. Status: 1=dead; 0=alive
6. DxtoB3: time from diagnosis to 1st BBBD in years
7. DxtoDeath: time from diagnosis to death in years
8. B3toDeath: time from 1st BBBD to death in years
9. KPS.PRE.: Karnofsky performance score before 1st BBBD, numerical value 0 − 100
10. LESSING: Lesions; single=0; multiple=1
11. LESDEEP: Lesions: superficial=0; deep=1
12. LESSUP: Lesions; supra=0; infra=1; both=2
13. PROC: Procedure; subtotal resection=1; biopsy=2; other=3
14. RAD4000: Radiation > 4000; yes=1; no=0
15. CHEMOPRIOR: yes=1; no=0
16. RESPONSE: Tumor response to chemo − complete=1; partial=2; blanks represent missing data

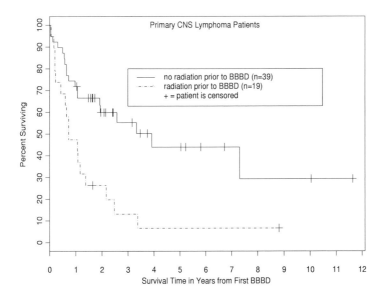

Figure 1.1 *Survival functions for CNS data.*

the average, is Group 0 (no prior radiation) surviving as long or longer with improved cognitive function? Figure 1.1 displays the two estimated survival curves. This type of nonparametric curve is defined in Chapter 2. Group 0's curve is always above that of Group 1 suggesting a higher rate of survival and hence a longer average survival time for Group 0. (It has been well documented in prior studies as well as in this study that radiation profoundly impairs cognitive functioning.) Further, is there a significant dependence of survival time, and/or the difference in survival, on any subset of the available covariates? That is, do any subset of the covariates help to explain survival time? For example, does age at time of first treatment or gender increase or decrease the relative risk of survival? Certainly we want to implement some kind of regression procedure that accommodates censored data and addresses the notions of relative risk and survival rates.

A naive descriptive analysis of AML study:

We consider a couple of descriptive measures to compare the two groups of data given in Example 1. The first approach is to throw out censored observations, the second is to treat the censored observations as exact ones, and the last is to use them all as they are. We at least expect to see different results among the three approaches. Let's see just how different they are.

• Analysis of AML data after throwing out censored observations

Measures	Maintained	Nonmaintained
Mean	25.1	21.7
Median	23.0	23.0

The mean for maintained group is slightly larger than that for nonmaintained group while their medians are the same. That is, the distribution of maintained group is slightly more skewed to the right than the nonmaintained group's distribution is. The difference between the two groups appears to be negligible.

• Analysis of AML data treating censored observations as exact

Measures	Maintained	Nonmaintained
Mean	38.5	21.3
Median	28.0	19.5

Both the mean and median for maintained group are larger than those for nonmaintained group. The difference between the two groups seems to be non-negligible in terms of both mean and median. The skewness of the maintained group is even more pronounced. We expect, however, that these estimates are biased in that they underestimate the true mean and median. The censored times are smaller than the true unknown failure times. The next analysis is done using a method which accommodates the censored data.

- Analysis of AML data accounting for the censoring

$Measures$	$Maintained$	$Nonmaintained$
$Mean$	52.6	22.7
$Median$	31.0	23.0

Both the mean and median for maintained group are larger than those for non-maintained group. Further, the mean of the maintained group is much larger than that of the nonmaintained group. Here we notice that the distribution of maintained group is much more skewed to the right than the nonmaintained group's distribution is. Consequently, the difference between the two groups seems to be huge. From this small example, we have learned that appropriate methods should be applied in order to deal with censored data. The method used here to estimate the mean and median is discussed in Chapter 2.1.

1.2 Basic definitions

Let T denote a nonnegative random variable representing the lifetimes of individuals in some population. ("Nonnegative" means $T \geq 0$.) We treat the case where T is continuous. For a treatment of discrete models see Lawless (1982, page 10). Let $F(\cdot)$ denote the (cumulative) **distribution function** (d.f.) of T with corresponding **probability density function** (p.d.f.) $f(\cdot)$. Note $f(t) = 0$ for $t < 0$. Then

$$F(t) = P(T \leq t) = \int_0^t f(x)dx. \tag{1.1}$$

The probability that an individual survives to time t is given by the **survivor function**

$$S(t) = P(T \geq t) = 1 - F(t) = \int_t^\infty f(x)dx. \tag{1.2}$$

This function is also referred to as the **reliability function**. Note that $S(t)$ is a monotone decreasing function with $S(0) = 1$ and $S(\infty) = \lim_{t \to \infty} S(t) = 0$.

Conversely, we can express the p.d.f. as

$$f(t) = \lim_{\Delta t \to 0^+} \frac{P(t \leq T < t + \Delta t)}{\Delta t} = \frac{dF(t)}{dt} = -\frac{dS(t)}{dt}. \tag{1.3}$$

The **pth-quantile** of the distribution of T is the value t_p such that

$$F(t_p) = P(T \leq t_p) = p. \tag{1.4}$$

That is, $t_p = F^{-1}(p)$. The pth-quantile is also referred to as the **100 × pth percentile** of the distribution. The **hazard function** specifies the instantaneous rate of failure at $T = t$ given that the individual survived up to time t and is defined as

$$h(t) = \lim_{\Delta t \to 0^+} \frac{P(t \leq T < t + \Delta t | T \geq t)}{\Delta t} = \frac{f(t)}{S(t)}. \tag{1.5}$$

We see here that $h(t)\Delta t$ is approximately the probability of a death in $[t, t + \Delta t)$, given survival up to time t. The hazard function is also referred to as the **risk** or **mortality rate**. We can view this as a measure of intensity at time t or a measure of the potential of failure at time t. The hazard is a rate, rather than a probability. It can assume values in $[0, \infty)$.

To understand why the hazard is a rate rather than a probability, in its definition consider the expression to the right of the limit sign which gives the ratio of two quantities. The numerator is a conditional probability and the denominator is Δt, which denotes a small time interval. By this division, we obtain a probability per unit time, which is no longer a probability but a rate. This ratio ranges between 0 and ∞. It depends on whether time is measured in days, weeks, months, or years, etc. The resulting value will give a different number depending on the units of time used. To illustrate this let $P = P(t \leq T < t + \Delta t | T \geq t) = 1/4$ and see the following table:

P	Δt	$\frac{P}{\Delta t}$ = rate
$\frac{1}{4}$	$\frac{1}{3}$ day	$\frac{1/4}{1/3} = 0.75$/day
$\frac{1}{4}$	$\frac{1}{21}$ week	$\frac{1/4}{1/21} = 5.25$/week

It is easily verified that $h(t)$ specifies the distribution of T, since

$$h(t) = -\frac{dS(t)/dt}{S(t)} = -\frac{d \log \left(S(t) \right)}{dt}.$$

Integrating $h(u)$ over $(0, t)$ gives the **cumulative hazard function** $H(t)$:

$$H(t) = \int_0^t h(u)du = -\log \left(S(t) \right). \tag{1.6}$$

In this book, unless otherwise specified, log denotes the natural logarithm, the inverse function of the exponential function $\exp = e$. Thus,

$$S(t) = \exp\left(-H(t)\right) = \exp\left(-\int_0^t h(u)du\right). \qquad (1.7)$$

Hence, the p.d.f. of T can be expressed as

$$f(t) = h(t)\exp\left(-\int_0^t h(u)du\right).$$

Note that $H(\infty) = \int_0^\infty h(t)dt = \infty$.

For a nonnegative random variable T **the mean value**, written $E(T) = \int_0^\infty t \cdot f(t)dt$, can be shown to be

$$E(T) = \int_0^\infty S(t)dt. \qquad (1.8)$$

WHY! Thus, mean survival time is the total area under the survivor curve $S(t)$. It follows from expression (1.7), for a given time t, the greater the risk, the smaller $S(t)$, and hence the shorter mean survival time $E(T)$, and vice versa. The following picture should help you to remember this relationship.

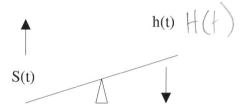

Another basic parameter of interest is the **mean residual life** at time u, denoted by $\mathrm{mrl}(u)$. For individuals of age u, this parameter measures their expected remaining lifetime. It is defined as

$$\mathrm{mrl}(u) = E(T - u \mid T > u).$$

For a continuous random variable it can be verified that

$$\mathrm{mrl}(u) = \frac{\int_u^\infty S(t)dt}{S(u)}. \qquad (1.9)$$

WHY! The $\mathrm{mrl}(u)$ is hence the area under the survival curve to the right of u divided by $S(u)$. Lastly, note the mean life, $E(T) = \mathrm{mrl}(0)$, is the total area under the survivor curve. The graph in Figure 1.2 illustrates this definition.

To end this section we discuss hazard functions and p.d.f.'s for three continuous distributions displayed in Figure 1.3. Model (a) has an increasing hazard rate. This may arise when there is a natural aging or wear. Model (b) has a decreasing hazard rate. Decreasing functions are less common but find occasional use when there is an elevated likelihood of early failure, such as in

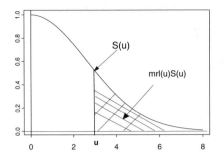

Figure 1.2 *Mean residual life at time u.*

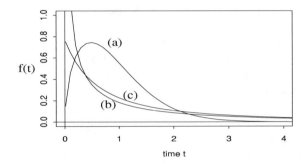

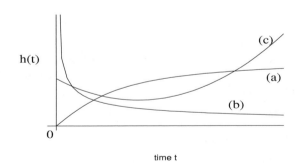

Figure 1.3 *Types of hazard rates and respective densities.*

certain types of electronic devices or in patients experiencing certain types of organ transplants. Model (c) has a bathtub-shaped hazard. Most often these are appropriate for populations followed from birth. Similarly, some manufactured equipment may experience early failure due to defective parts, followed by a constant hazard rate which, in later stages of equipment life, increases. Most population mortality data follow this type of hazard function where, during an early period, deaths result, primarily from infant diseases, after which the death rate stabilizes, followed by an increasing hazard rate due to the natural aging process. Not represented in these plots is the hump-shaped hazard; i.e., the hazard is increasing early and then eventually begins declining. This type of hazard rate is often used to model survival after successful surgery where there is an initial increase in risk due to infection, hemorrhaging, or other complications just after the procedure, followed by a steady decline in risk as the patient recovers.

Remark:

Although different survivor functions can have the same basic shape, their hazard functions can differ dramatically, as is the case with the previous three models. The hazard function is usually more informative about the underlying mechanism of failure than the survivor function. For this reason, modelling the hazard function is an important method for summarizing survival data.

1.3 Censoring and truncation models

We now present six types of censoring models and two truncation models. Let $T_1, T_2, \ldots, T_n$ be independent and identically distributed (iid) with distribution function (d.f.) F.

Type I censoring

This type arises in engineering applications. In such situations there are transistors, tubes, chips, etc.; we put them all on test at time $t = 0$ and record their times to failure. Some items may take a long time to "burn out" and we will not want to wait that long to terminate the experiment. Therefore, we terminate the experiment at a prespecified time t_c. The number of observed failure times is random. If n is the number of items put on test, then we could observe $0, 1, 2, \ldots, n$ failure times. The following illustrates a possible trial:

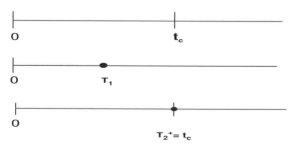

We call t_c the fixed censoring time. Instead of observing the T_i, we observe $Y_1, Y_2, \ldots, Y_n$ where

$$Y_i = \min(T_i, t_c) = \begin{cases} T_i & \text{if } T_i \leq t_c \\ t_c & \text{if } t_c < T_i. \end{cases}$$

Notice that the d.f. of Y has positive mass $P(T > t_c) > 0$ at $y = t_c$ since the $P(Y = t_c) = P(t_c < T) = 1 - F(t_c) > 0$. That is, Y is a mixed random variable with a continuous and discrete component. The (cumulative) d.f. $M(y)$ of Y is shown in Figure 1.4. It is useful to introduce a binary random variable δ which indicates if a failure time is observed or censored,

$$\delta = \begin{cases} 1 & \text{if } T \le t_c \\ 0 & \text{if } t_c < T. \end{cases}$$

Note that ($\delta = 0$ and $T \le t_c$) implies that the failure time was precisely $T = t_c$, which occurs with zero probability if T is a continuous variable. (Note that for discrete distributions, we can set t_c equal to the last attainable time a failure may be observed. Hence, the probability $P(\{\delta = 0\} \cap \{T \le t_c\})$ is not equal to zero.) We then observe the iid random pairs (Y_i, δ_i).

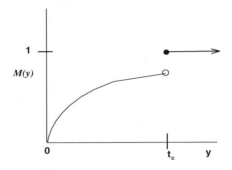

Figure 1.4 *Cumulative d.f. of the mixed random variable Y.*

For <u>maximum likelihood estimation</u> (detailed in Chapter 3.2) of any parameters of the distribution of T, we need to calculate the joint likelihood of the pair (Y, δ). By <u>likelihood</u> we mean the rubric which regards the density as a function of the parameter for a given (fixed) value (y, δ). For $y < t_c$, $P(Y \le y) = P(T \le y) = F(y)$ and $P(\delta = 1 \mid Y \le y) = 1$. Therefore, the likelihood for $Y = y < t_c$ and $\delta = 1$ is the density $f(y)$. For $y = t_c$ and $\delta = 0$, the likelihood for this event is the probability $P(\delta = 0, Y = t_c) = P(T > t_c) = S(t_c)$.

We can combine these two expressions into one single expression $f(y)^{\delta} \times S(t_c)^{1-\delta}$. As usual, we define the likelihood function of a random sample to be the product of the densities of the individual observations. That is, the likelihood function for the n iid random pairs (Y_i, δ_i) is given by

$$L = \prod_{i=1}^{n} f(y_i)^{\delta_i} S(t_c)^{1-\delta_i}. \tag{1.10}$$

Type II censoring

In similar engineering applications as above, the censoring time may be left open at the beginning. Instead, the experiment is run until a prespecified fraction r/n of the n items has failed. Let $T_{(1)}, T_{(2)}, \ldots, T_{(n)}$ denote the ordered values of the random sample $T_1, \ldots, T_n$. By plan, <u>observations terminate after the rth failure occurs</u>. So we only observe the r smallest observations in a random sample of n items. For example, let $n = 25$ and take $r = 15$. Hence, when we observe 15 burn out times, we terminate the experiment. Notice that we could wait an arbitrarily long time to observe the 15th failure time as $T_{(15)}$ is random. The following illustrates a possible trial:

In this trial the last 10 observations are assigned the value of $T_{(15)}$. Hence we have 10 censored observations. More formally, we observe the following full sample.

$$
\begin{aligned}
Y_{(1)} &= T_{(1)} \\
Y_{(2)} &= T_{(2)} \\
&\vdots \\
Y_{(r)} &= T_{(r)} \\
Y_{(r+1)} &= T_{(r)} \\
&\vdots \\
Y_{(n)} &= T_{(r)}.
\end{aligned}
$$

Formally, the data consist of the r smallest lifetimes $T_{(1)}, \ldots, T_{(r)}$ out of the n iid lifetimes $T_1, \ldots, T_n$ with continuous p.d.f $f(t)$ and survivor function $S(t)$. Then the likelihood function (joint p.d.f) of $T_{(1)}, \ldots, T_{(r)}$ is given

$$ L = \frac{n!}{(n-r)!} f(t_{(1)}) \cdots f(t_{(r)}) \Big(S(t_{(r)}) \Big)^{n-r}. \tag{1.11} $$

WHY!

Remarks:

1. In Type I censoring, the endpoint t_c is a fixed value and the number of observed failure times is a random variable which assumes a value in the set $\{0, 1, 2, \ldots, n\}$.

2. In Type II censoring, the number of failure times r is a fixed value whereas the endpoint T_r is a random observation. Hence we could wait possibly a very long time to observe the r failures or, vice versa, see all r relatively early on.

3. Although Type I and Type II censoring are very different designs, **the form of the observed likelihood function is the same in both cases**. To see this it is only necessary to note that the individual items whose lifetimes are observed contribute a term $f(y_{(i)})$ to the observed likelihood function,

whereas items whose lifetimes are censored contribute a term $S(y_{(i)})$. The factor $n!/(n-r)!$ in the last equation reflects the fact that we consider the ordered observations. For maximum likelihood estimation the factor will be irrelevant since it does not depend on any parameters of the distribution function.

Random censoring

Right censoring is presented here. Left censoring is analogous. Random censoring occurs frequently in medical studies. In clinical trials, patients typically enter a study at different times. Then each is treated with one of several possible therapies. We want to observe their "failure" time but censoring can occur in one of the following ways:

1. *Loss to Follow-up*. Patient moves away. We never see him again. We only know he has survived from entry date until he left. So his survival time is $\geq$ the observed value.

2. *Drop Out*. Bad side effects forces termination of treatment. Or patient refuses to continue treatment for whatever reasons.

3. *Termination of Study*. Patient is still "alive" at end of study.

The following illustrates a possible trial:

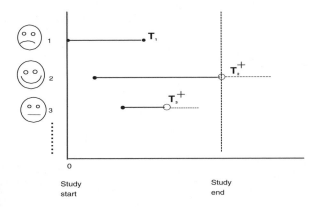

Here, patient 1 entered the study at $t = 0$ and died at time T_1 to give an uncensored observation; patient 2 entered the study, and by the end of the study he was still alive resulting in a censored observation T_2^+; and patient 3 entered the study and was lost to follow-up before the end of the study to give another censored observation T_3^+. The AML and CNS lymphoma studies in Examples 1 and 2 contain randomly right-censored data.

Let T denote a lifetime with d.f. F and survivor function S_f and C denote a random censor time with d.f. G, p.d.f. g, and survivor function S_g. Each individual has a lifetime T_i and a censor time C_i. On each of n individuals we observe the pair (Y_i, δ_i) where

$$Y_i = \min(T_i, C_i) \quad \text{and} \quad \delta_i = \begin{cases} 1 & \text{if } T_i \le C_i \\ 0 & \text{if } C_i < T_i . \end{cases}$$

Hence we observe n iid random pairs (Y_i, δ_i). The times T_i and C_i are usually assumed to be independent. This is a strong assumption. If a patient drops out because of complications with the treatment (case 2 above), it is clearly offended. However, under the independence assumption, the likelihood function has a simple form (1.12), and even simpler in expression (1.13). Otherwise, we lose the simplicity. The likelihood function becomes very complicated and, hence, the analysis is more difficult to carry out.

Let M and S_m denote the distribution and survivor functions of $Y = \min(T, C)$ respectively. Then by the independence assumption it easily follows that the survivor function is

$$S_m(y) = P(Y > y) = P(T > y, C > y) = P(T > y)P(C > y) = S_f(y)S_g(y).$$

The d.f. of Y is $M(y) = 1 - S_f(y)S_g(y)$.

The likelihood function of the n iid pairs (Y_i, δ_i) is given by

$$\begin{aligned} L &= \prod_{i=1}^{n} \Big(f(y_i)S_g(y_i) \Big)^{\delta_i} \cdot \Big(g(y_i)S_f(y_i) \Big)^{1-\delta_i} \\ &= \left(\prod_{i=1}^{n} S_g(y_i)^{\delta_i} g(y_i)^{1-\delta_i} \right) \cdot \left(\prod_{i=1}^{n} f(y_i)^{\delta_i} S_f(y_i)^{1-\delta_i} \right). \end{aligned} \tag{1.12}$$

Note: If the distribution of C does not involve any parameters of interest, then the first factor plays no role in the maximization process. Hence, the likelihood function can be taken to be

$$L = \prod_{i=1}^{n} f(y_i)^{\delta_i} S_f(y_i)^{1-\delta_i}, \tag{1.13}$$

which has the same form as the likelihood derived for both Type I (1.10) and Type II (1.11) censoring. Thus, regardless of which of the three types of censoring is present, the maximization process yields the same estimated quantities.

The derivation of the likelihood is as follows:

$$\begin{aligned} P(Y = y, \delta = 0) &= P(C = y, C < T) = P(C = y, y < T) \\ &= P(C = y)P(y < T) \quad \text{by independence} \\ &= g(y)S_f(y). \\ P(Y = y, \delta = 1) &= P(T = y, T < C) = P(T = y, y < C) = f(y)S_g(y) . \end{aligned}$$

Hence, the joint p.d.f. of the pair (Y, δ) (a mixed distribution as Y is continuous and δ is discrete) is given by the single expression

$$P(y, \delta) = \Big(g(y)S_f(y)\Big)^{1-\delta} \cdot \Big(f(y)S_g(y)\Big)^{\delta}.$$

The likelihood of the n iid pairs (Y_i, δ_i) given above follows.

Case 1 interval censored data: current status data

Consider the following two examples which illustrate how this type of censoring arises.

Example 3.

Tumor free laboratory mice are injected with a tumor inducing agent. The mouse must be killed in order to see if a lung tumor was induced. So after a random period of time U for each mouse, it is killed and the experimenter checks to see whether or not a tumor developed. The endpoint of interest is T, "time to tumor."

Example 4.

An ophthalmologist developed a new treatment for a particular eye disease. To test its effectiveness he must conduct a clinical trial on people. His endpoint of interest is "time to cure the disease." We see this trial could produce right-censored data. During the course of this study he notices an adverse side effect which impairs vision in some of the patients. So now he wants to study "time to side effect" where he has a control group to compare to the treatment group to determine if this impairment is indeed due to the new treatment. Let's focus on the treatment group. All these patients received the new treatment. In order to determine "time to side effect" T, he takes a snapshot view. At a random point in time he checks all patients to see if they developed the side effect. The records ministry keeps very precise data on when each patient received the new treatment for the disease. So the doctor can look back in time from where he takes his snapshot to the time of first treatment. Hence, for each patient we have an observed time U, which equals time from receiving new treatment to the time of the snapshot. If the patient has the side effect, then his $T \leq U$. If the patient is still free of the side effect, then his $T > U$.

In both examples the only available observed time is the U, the censoring time. The following illustrates a possible trial of Example 3.

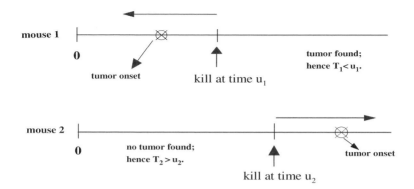

More formally, we observe only the iid times $U_i, i = 1, \ldots, n$ and $\delta_i = I\{T_i \leq U_i\}$. That is, $\delta = 1$ if the event $T \leq U$ has occurred, and $\delta = 0$ if the event has not occurred. We assume the **support** (the interval over which the distribution has positive probability) of U is contained in the support of T. As before, the $T \sim F$ and the censor time $U \sim G$ and again we assume T and U are independent random times. The derivation of the joint p.d.f. of the pair of (U, δ) follows:

$$P(U = u, \delta = 0) = P(\delta = 0|U = u)P(U = u)$$
$$= P(T > u)P(U = u) = S_f(u)g(u).$$
$$P(U = u, \delta = 1) = P(\delta = 1|U = u)P(U = u)$$
$$= P(T \leq u)P(U = u) = F(u)g(u).$$

We can write this joint p.d.f. of the pair (U, δ) (again a mixed distribution) in a single expression

$$P(u, \delta) = \Big(S_f(u)\Big)^{1-\delta} \Big(F(u)\Big)^{\delta} g(u).$$

The likelihood of the n iid pairs (U_i, δ_i) easily follows.

Left-censored and doubly-censored data

The following two examples illustrate studies where left-censored, uncensored, and right-censored observations could occur. When all these can occur, this is often referred to as doubly-censored data.

Example 5.

A child psychiatrist visits a Peruvian village to study the age at which children first learn to perform a particular task. Let T denote the age a child learns to perform a specified task. The following picture illustrates the possible outcomes:

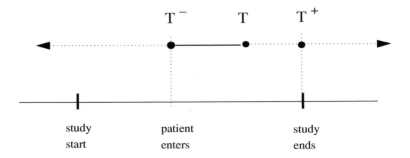

We read the recorded values as follows: T: exact age is observed (uncensored), T^-: age is left-censored as the child already knew the task when s/he was initially tested in the study, and T^+: age is right-censored since the child did not learn the task during the study period.

Example 6.

Extracted from Klein & Moeschberger (1997): High school boys are interviewed to determine the distribution of the age of boys when they first used marijuana. The question stated was "When did you first use marijuana?" The three possible answers and respective recorded values are given in the following table:

Possible answer:	Recorded value:
a. I used it but I cannot recall just when the first time was.	a. T^-: age at interview as exact age was earlier but unknown
b. I first used it when I was ___.	b. T: exact age since it is known (uncensored)
c. I never used it.	c. T^+: age at interview since exact age occurs sometime in the future

Interval censoring

The time-to-event T is known only to occur within an interval. Such censoring occurs when patients in clinical trial or longitudinal study have *periodic* follow-up. For example, women in a study are required to have yearly PAP smear exams. Each patient's event time T_i is only known to fall in an interval $(L_i, R_i]$ which represents the time interval between the visit prior to the visit when the event of interest is detected. The L_i and R_i denote respectively the left and right endpoints of the censoring interval. For example, if the ith patient shows the sign of the symptom at her first follow-up time, then L_i is zero; in other words, the origin of the study and R_i is her first follow-up time. Further, if she showed no sign of the symptom until her $i-1$th follow-up times but shows the sign of the symptom at her ith follow-up, then L_i is her $i-1$th follow-up and R_i is her ith follow-up. If she doesn't exhibit the symptom at her last follow-up, L_i is her last follow-up and R_i is ∞. Note that any combination of left, right, or interval censoring may occur in a study. Furthermore, we see that left censoring, right censoring, and current status data are special cases of interval censoring.

Truncation

Here we summarize Klein & Moeschberger's (1997, Sections 1.16, 1.19, and 3.4) discussion of truncation. Truncation is a procedure where a condition other than the main event of interest is used to screen patients; that is, only if the patient has the truncation condition prior to the event of interest will s/he be observed by the investigator. Hence, there will be subjects "rejected" from the study so that the investigator will never be aware of their existence. This truncation condition may be exposure to a certain disease, entry into a retirement home, or an occurrence of an intermediate event prior to death. In this case, the main event of interest is said to be *left-truncated*. Let U denote the time at which the truncation event occurs and let T denote the time of the main event of interest to occur. Then for left-truncated samples, only individuals with $T \geq U$ are observed. The most common type of *left truncation* occurs when subjects enter the study at a random age and are followed from this *delayed entry time* until the event of interest occurs or the subject is right-censored. In this situation, all subjects who experience the event of interest prior to the delayed entry time will not be known to the experimenter. The following example of *left-truncated and right-censored* data is described in Klein & Moeschberger (1997, pages 15−17, and Example 3.8, page 65). In Chapter 7.3 we treat the analysis of left-truncated and right-censored data.

Example 7. Death times of elderly residents of a retirement community

Age in months when members of a retirement community died or left the center (right-censored) and age when the members entered the community (the truncation event) are recorded. Individuals must survive to a sufficient age to enter the retirement community. Individuals who die at an early age are excluded from the study. Hence, the life lengths in this data set are *left-truncated*. Ignoring this truncation leads to problem of *length-biased sampling*. We want a survival analysis to account for this type of bias.

Right truncation occurs when only individuals who have experienced the main event of interest are included in the sample. All others are excluded. A mortality study based on death records is a good example of this. The following example of *right-truncated* data is described in Klein & Moeschberger (1997, page 19, and Example 3.9, page 65).

Example 8. Time to AIDS

Measurement of interest is the waiting time in years from HIV infection to development of AIDS. In the sampling scheme, only individuals who have developed AIDS prior to the end of the study are included in the study. Infected individuals who have yet to develop AIDS are excluded from the sample; hence, unknown to the investigator. This is a case of *right truncation*.

1.4 Course objectives

The objectives here are to learn methods to model and analyze the data like those presented in the two examples in Section 1.1. We want these statistical procedures to accommodate censored data and to help us attain **the three basic goals of survival analysis** as so succinctly delineated by Kleinbaum (1995, page 15).

In Table 1.3, the graph for **Goal 1** illustrates the survivor functions give very different interpretations. The left one shows a quick drop in survival probabilities early in follow-up. Then the rate of decrease levels off later on. The right function, in contrast, shows a very slow decrease for quite a long while, then a sharp decrease much later on.

In Table 1.3, the plot for **Goal 2** shows that up to 13 weeks, the graph for the new method lies above that for the old. Thereafter the graph for old method is above the new. Hence, this dual graph reveals that up to 13 weeks the new method is more effective than the old; however, after 13 weeks, it becomes less effective.

In Table 1.3, the graph for **Goal 3** displays that, for any fixed point in time, up to about 10 years of age, women are at greater risk to get the disease than men are. From 10 to about 40 years of age, men now have a slightly greater risk. For both genders the hazard function decreases as the person ages.

Remark:

As usual, the emphasis is on modelling and inference. Modelling the hazard function or failure time in turn provides us with estimates of population features such as the mean, the mean residual life, quantiles, and survival probabilities.

Table 1.3:   *Goals of survival analysis*

Goal 1. To estimate and interpret survivor and/or hazard functions from survival data.

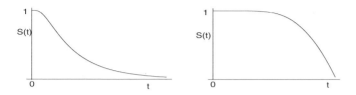

Goal 2. To compare survivor and/or hazard functions.

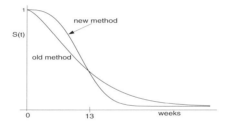

Goal 3. To assess the relationship of explanatory variables to survival time, especially through the use of formal mathematical modelling.

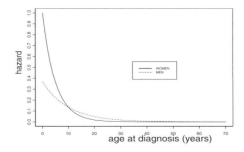

1.5 Data entry and import/export of data files

The layout is a typical spreadsheet format which is virtually the same for all data analytic software packages. Some examples are EXCEL, SPSS, MINITAB, SAS. The spreadsheet in S-PLUS is the data object called a `data.frame`. On the standard toolbar menu click sequentially on the white blank page at upper far left, `File → New → Data Set → Ok`. A new (empty) `data.frame` will appear. This likens an EXCEL spreadsheet. Double right click on the cell just below the column number to enter the variable name. Below is a table which displays our S-PLUS data set "aml.data" along with a key. This `data.frame` object contains the AML data first given in Table 1.1 under Example 1, page 2. Note that **status variable = the indicator variable δ**. This data set is saved as, e.g., "aml.sdd." You can also save this data set as an Excel file. Just click on `File → ExportData → ToFile`. Go to `Save as` and click `Type → MicrosoftExcelFiles` (*.xls).

	1	2	3	
	weeks	group	status	
1	9	1	1	
2	13	1	1	
3	13	1	0	**group** = 1 for maintained,
4	18	1	1	**group** = 0 for nonmaintained.
.	.	.	.	
.	.	.	.	**status** = 1 if uncensored
.	.	.	.	(relapse occurred),
11	161	1	0	**status** = 0 if censored (still in
12	5	0	1	remission; recorded with + sign).
13	5	0	1	
14	8	0	1	
.	.	.	.	
.	.	.	.	
.	.	.	.	
23	45	0	1	

It seems that EXCEL has spread itself worldwide. All the mainstream statistical packages can accept an EXCEL file. Feel free to first enter your data in an EXCEL spreadsheet. To import into S-PLUS do the following sequentially: in S-PLUS, click on File → ImportData → FromFile → FilesofType → MicrosoftExcelFiles (*.xl*). In Look In, find your way to the directory where your desired *.xls data file is. Then right-click on it and click on Open. It's now in an S-PLUS data sheet. You can save it in S-PLUS as an S-PLUS data file (data.frame object). Click on File, then on Save. It should be clear from this point. Your file will be saved as a *.sdd file.

To import your data file into S or R, first save your EXCEL file, or any other file, as a *.txt file. Be sure to open this file first to see what the delimiter is; that is, what is used to separate the data values entered on each row. Suppose your data file, called your.txt, is in the C: directory. The S and R function read.table imports your.txt file and creates a data.frame object. When a comma is the delimiter, use the following S line command:

> your <- read.table("C://your.txt",header = T,sep = ",")

for tabbed use sep = "\t"

If the delimiter is "~", use sep = "~". If blank space separates the data values, use sep = " ". In R, to perform a survival analysis it is necessary to install the survival analysis library. The R command is

> library(survival)

The R function require(survival) accomplishes the same.

1.6 Exercises

A. *Applications*

Identify the data types of the following cases:

1.1 Suppose that six rats have been exposed to carcinogens by injecting tumor cells into their foot-pads. The times to develop a tumor of a given size are observed. The investigator decides to terminate the experiment after 30 weeks. Rats A, B, and D develop tumors after 10, 15, and 25 weeks, respectively. Rats C and E do not develop by the end of the study. Rat F died accidentally without any tumors after 19 weeks of observation. (*Source*: Lee, E.T. (1992, page 2). *Statistical Methods for Survival Data Analysis, 2nd ed.*, New York: John Wiley & Sons.)

1.2 In Exercise 1.1, the investigator may decide to terminate the study after four of the six rats have developed tumors. Rats A, B, and D develop tumors after 10, 15, and 25 weeks, respectively. Rat F died accidentally without any tumors after 19 weeks of observation. Rat E develops tumor after 35 weeks but Rats C does not develop by that time. How would the data set in Exercise 1.1 change? (*Source*: Lee, E.T. (1992, pages 2−3). *Statistical*

Methods for Survival Data Analysis, 2nd ed., New York: John Wiley & Sons.)

1.3 Suppose that six patients with acute leukemia enter a clinical study during a total study period of one year. Suppose also that all six respond to treatment and achieve remission. Patients A, C, and E achieve remission at the beginning of the second, fourth, and ninth months and relapse after four, six, and three months, respectively. Patient B achieves remission at the beginning of the third month but is lost to follow-up four months later. Patients D and F achieve remission at the beginning of the fifth and tenth month, respectively, and are still in remission at the end of the study. Find out the remission times of the six patients. (*Source*: Lee, E.T. (1992, pages 3–4). *Statistical Methods for Survival Data Analysis, 2nd ed.*, New York: John Wiley & Sons.)

1.4 Survival/sacrifice experiments are designed to determine whether a suspected agent accelerates the time until tumor onset in experimental animals. For such studies, each animal is assigned to a prespecified dose of a suspected carcinogen, and examined at sacrifice or death, for the presence or absence of a tumor. Since a lung tumor is occult, the time until tumor onset is not directly observable. Instead, we observe only a time of sacrifice or death. (*Source*: Hoel, D.G. and Walburg, H.E., Jr. (1972). Statistical analysis of survival experiments. *J. Natl. Cancer Inst.*, **49**, 361 − 372.)

1.5 An annual survey on 196 girls recorded whether or not, at the time of the survey, sexual maturity had developed. Development was complete in some girls before the first survey, some girls were lost before the last survey and before development was complete, and some girls had not completed development at the last survey. (*Source*: Peto, R. (1973). Empirical survival curves for interval censored data. *Appl. Statist.*, **22**, 86 − 91.)

1.6 Woolson (1981) has reported survival data on 26 psychiatric inpatients admitted to the University of Iowa hospitals during the years 1935 − 1948. This sample is part of a larger study of psychiatric inpatients discussed by Tsuang and Woolson (1977). Data for each patient consists of age at first admission to the hospital, sex, number of years of follow-up (years from admission to death or censoring), and patient status at the follow-up time. The main goal is to compare the survival experience of these 26 patients to the standard mortality of residents of Iowa to determine if psychiatric patients tend to have shorter lifetimes. (*Source*: Klein, J.P. and Moeschberger, M.L. (1997, page 15). *Survival Analysis: Techniques for Censored and Truncated Data*. New York: Springer.)

1.7 The US Centers for Disease Control maintains a database of reported AIDS cases. We consider the 1,927 cases who were infected by contaminated blood transfusions and developed AIDS by November 1989. For our data, the earliest reported infection date was January 1975. For our analysis, we give a code of 0 for young children (ages 0 − 12) and 1 for older children and adults (ages 13 and up). We wish to test whether the induction periods for the two groups have the same latency distribution. (*Source*: Finkelstein, D.M., Moore, D.F., and Schoenfeld, D.A. (1993). A proportional hazards model for truncated AIDS data. *Biometrics*, **49**, 731 − 740.)

1.8 Leiderman *et al.* wanted to establish norms for infant development for a community in Kenya in order to make comparisons with known standards in the United States and the United Kingdom. The sample consisted of 65 children born between July 1 and December 31, 1969. Starting in January 1970, each child was tested monthly to see if he had learned to accomplish certain standard tasks. Here the variable of interest T would represent the time from birth to first learn to perform a particular task. Late entries occurred when it was found that, at the very first test, some children could already perform the task, whereas losses occurred when some infants were still unsuccessful by the end of the study. (*Source*: Leiderman, P.H., Babu, D., Kagia, J., Kraemer, H.C., and Leiderman, G.F. (1973). African infant precocity and some social influences during the first year. *Nature*, **242**, 247 − 249.)

B. *Theory and WHY!*

1.9 Show expression (1.8).
   Hint: Use $E(T) = \int_0^\infty \left(\int_0^t dx \right) f(t)dt$.

.10 Verify expression (1.9).
   Hint: Examine expressions (1.3) and (1.5).

.11 Derive expression (1.11).
   Hint: Refer to the **Remark** in Hogg and Craig (1995, pages 199 − 200).

CHAPTER 2

Nonparametric Methods

We begin with nonparametric methods of inference concerning the survivor function $S(t) = P(T > t)$ and, hence, functions of it.

Objectives of this chapter:

After studying Chapter 2, the student should:

1. Know how to compute the **Kaplan-Meier** (K-M) estimate of survival and **Greenwood's** estimate of asymptotic variance of K-M at time t.

2. Know how to use the **redistribute-to-the-right algorithm** to compute the K-M estimate.

3. Know how to estimate the hazard and cumulative hazard functions.

4. Know how to estimate the pth-quantile.

5. Know how to plot the K-M curve over time t in S.

6. Know how to implement the S function `survfit` to conduct nonparamtric analyses.

7. Know how to plot two K-M curves to compare survival between two (treatment) groups.

8. Be familiar with **Fisher's exact test**.

9. Know how to compute the **log-rank test statistic**.

10. Know how to implement the S function `survdiff` to conduct the log-rank test.

11. Understand why we might **stratify** and how this affects the comparison of two survival curves.

12. Understand how the log-rank test statistic is computed when we stratify on a covariate.

2.1 Kaplan-Meier estimator of survival

We consider the AML data again introduced in Table 1.1, Chapter 1.1. The ordered data is included here in Table 2.1 for ease of discussion.

Table 2.1: *Data for the AML maintenance study*

Group	Length of complete remission(in weeks)
Maintained	9, 13, 13+, 18, 23, 28+, 31, 34, 45+, 48, 161+
Nonmaintained	5, 5, 8, 8, 12, 16+, 23, 27, 30, 33, 43, 45

A + indicates a censored value.

We first treat this data as if there were NO censored observations. Let t_i denote an ordered observed value. The **empirical survivor function (esf)**, denoted by $S_n(t)$, is defined to be

$$S_n(t) = \frac{\#\text{ of observations} > t}{n} = \frac{\#\{t_i > t\}}{n}. \qquad (2.1)$$

The $S_n(t)$ is the proportion of patients still in remission after t weeks. Let's consider the AML maintained group data (AML1) on a time line:

```
|———|—|—|—|—|—|———|—|————————|—
0   9  13 18 23 28 31 34   45 48        161
```

The values of the **esf** on the maintained group are:

t	0	9	13	18	23	28	31	34	45	48	161
$S_n(t)$	$\frac{11}{11}$	$\frac{10}{11}$	$\frac{8}{11}$	$\frac{7}{11}$	$\frac{6}{11}$	$\frac{5}{11}$	$\frac{4}{11}$	$\frac{3}{11}$	$\frac{2}{11}$	$\frac{1}{11}$	0

The plot of this **esf** function in Figure 2.1 can be obtained by the following S commands. Here status is an 11×1 vector of 1's since we are ignoring that four points are censored. We store the AML data in a data frame called aml. The S function `survfit` calculates the $S_n(t)$ values.

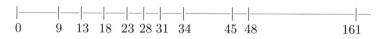

```
> aml1 <- aml[aml$group==1]
> status <- rep(1,11)
> esf.fit <- survfit(Surv(aml1,status)~1)
> plot(esf.fit,conf.int=F,xlab="time until relapse (in weeks)",
         ylab="proportion without relapse",lab=c(10,10,7))
> mtext("The Empirical Survivor Function of the AML Data",3,-3)
> legend(75,.80,c("maintained group","assuming no censored
   data"))
> abline(h=0)
```

The estimated median is the first value t_i where the $S_n(t) \leq 0.5$. Here the

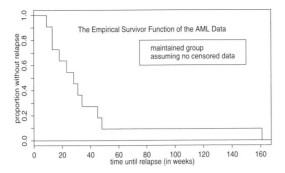

Figure 2.1 *Empirical survivor function (esf).*

$\widehat{\text{med}} = 28$ weeks. The estimated mean (expected value) is

$$\widehat{\text{mean}} = \int_0^\infty S_n(t)\, dt = \text{area under } S_n(t) = \bar{t}.$$

$S_n(t)$ is a right continuous step function which steps down at each t_i. The estimated mean then is just the sum of the areas of the ten rectangles on the plot. This sum is simply the sample mean. Here the $\widehat{\text{mean}} = \bar{t} = 423/11 = 38.45$ weeks.

Note: The **esf** is a consistent estimator of the true survivor function $S(t)$. The exact distribution of $nS_n(t)$, for each fixed t, is binomial (n, p), where $n =$ the number of observations and $p = P(T > t)$. Further, it follows from the central limit theorem that for each fixed t,

$$S_n(t) \quad \overset{a}{\sim} \quad \text{normal}(p, p(1 - p)/n),$$

where $\overset{a}{\sim}$ is read "approximately distributed as."

We now present the product-limit estimator of survival. This is commonly called the **Kaplan-Meier (K-M) estimator** as it appeared in a seminal 1958 paper.

The Product-limit (PL) estimator of $S(t) = P(T > t)$:

K-M adjusts the esf to reflect the presence of right-censored observations.

Recall the random right censoring model in Chapter 1.3. On each of n individuals we observe the pair (Y_i, δ_i) where

$$Y_i = \min(T_i, C_i) \quad \text{and} \quad \delta_i = \begin{cases} 1 & \text{if } T_i \leq C_i \\ 0 & \text{if } C_i < T_i. \end{cases}$$

On a time line we have

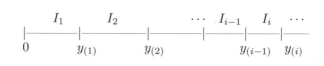

where $y_{(i)}$ denotes the ith distinct ordered censored or uncensored observation and is the right endpoint of the interval I_i, $i = 1, 2, \ldots, n' \leq n$.

- **death** is the generic word for the event of interest.
  In the AML study, a "relapse" (end of remission period) = "death"
- A **cohort** is a group of people who are followed throughout the course of the study.
- The people at risk at the beginning of the interval I_i are those people who survived (not dead, lost, or withdrawn) the previous interval I_{i-1} .
  Let $\mathcal{R}(t)$ denote the **risk set just before time t** and let

$$
\begin{aligned}
n_i &= \quad \# \text{ in } \mathcal{R}(y_{(i)}) \\
&= \quad \# \text{ alive (and not censored) just before } y_{(i)} \\
d_i &= \quad \# \text{ died at time } y_{(i)} \\
p_i &= \quad P(\text{surviving through } I_i \mid \text{alive at beginning } I_i) \\
&= \quad P(T > y_{(i)} \mid T > y_{(i-1)}) \\
q_i &= \quad 1 - p_i \quad = P(\text{die in } I_i \mid \text{alive at beginning } I_i).
\end{aligned}
$$

Recall the general multiplication rule for joint events A_1 and A_2:

$$P(A_1 \cap A_2) = P(A_2 \mid A_1)P(A_1).$$

From repeated application of this product rule the survivor function can be expressed as

$$S(t) = P(T > t) = \prod_{y_{(i)} \leq t} p_i.$$

The estimates of p_i and q_i are

$$\widehat{q}_i = \frac{d_i}{n_i} \quad \text{and} \quad \widehat{p}_i = 1 - \widehat{q}_i = 1 - \frac{d_i}{n_i} = \left(\frac{n_i - d_i}{n_i} \right).$$

The **K-M estimator of the survivor function** is

$$\widehat{S}(t) = \prod_{y_{(i)} \leq t} \widehat{p}_i = \prod_{y_{(i)} \leq t} \left(\frac{n_i - d_i}{n_i} \right) = \prod_{i=1}^{k} \left(\frac{n_i - d_i}{n_i} \right), \tag{2.2}$$

where $y_{(k)} \leq t < y_{(k+1)}$.

Let's consider the AML1 data on a time line where a "+" denotes a right-censored observed value. The censored time 13+ we place to the right of the observed relapse time 13 since the censored patient at 13 weeks was still in remission. Hence, his relapse time (if it occurs) is greater than 13 weeks.

$$
\begin{aligned}
\widehat{S}(0) &= 1 \\
\widehat{S}(9) &= \widehat{S}(0) \times \tfrac{11-1}{11} &= .91 \\
\widehat{S}(13) &= \widehat{S}(9) \times \tfrac{10-1}{10} &= .82 \\
\widehat{S}(13+) &= \widehat{S}(13) \times \tfrac{9-0}{9} &= \widehat{S}(13) &= .82 \\
\widehat{S}(18) &= \widehat{S}(13) \times \tfrac{8-1}{8} &= .72 \\
\widehat{S}(23) &= \widehat{S}(18) \times \tfrac{7-1}{7} &= .61 \\
\widehat{S}(28+) &= \widehat{S}(23) \times \tfrac{6-0}{6} &= \widehat{S}(23) &= .61 \\
\widehat{S}(31) &= \widehat{S}(23) \times \tfrac{5-1}{5} &= .49 \\
\widehat{S}(34) &= \widehat{S}(31) \times \tfrac{4-1}{4} &= .37 \\
\widehat{S}(45+) &= \widehat{S}(34) \times \tfrac{3-0}{3} &= \widehat{S}(34) &= .37 \\
\widehat{S}(48) &= \widehat{S}(34) \times \tfrac{2-1}{2} &= .18 \\
\widehat{S}(161+) &= \widehat{S}(48) \times \tfrac{1-0}{1} &= \widehat{S}(48) &= .18
\end{aligned}
$$

The K-M curve is a right continuous step function which steps down only at an uncensored observation. A plot of this together with the **esf** curve is displayed in Figure 2.2. The "+" on the K-M curve represents the survival probability at a censored time. Note the difference in the two curves. K-M is

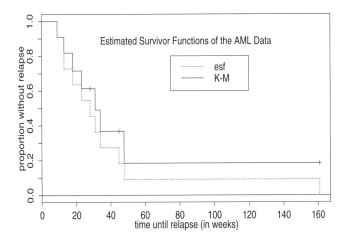

Figure 2.2 *Kaplan-Meier and esf estimates of survival.*

always greater than or equal to **esf**. When there are no censored data values

K-M reduces to the **esf**. Note the K-M curve does not jump down to zero as the largest survival time (161^{+}) is censored. We cannot estimate $S(t)$ beyond $t = 48$. Some refer to $\widehat{S}(t)$ as a defective survival function. Alternatively, $\widehat{F}(t) = 1 - \widehat{S}(t)$ is called a subdistribution function as the total probability is less than one.

We now describe an alternate, but equivalent, way to compute the K-M estimate of survival beyond time point t. Efron (1967) called this method "redistribute-to-the-right algorithm." This algorithm allows us to calculate K-M in a similar manner to **esf**. We adopt the interpretation provided by Gooley *et al.* (1999, 2000).

Redistribute-to-the-right algorithm:

We assume all patients are equally likely to fail. Hence, if all patients have either failed or been followed to a specified time point t, then each patient contributes a prescribed and equal amount to the estimate of the probability of failure:

$$\frac{1}{\mathcal{N}(t)},$$

where $\mathcal{N}(t) = $ total # of patients under study at or before t.

Each patient under study has a potential contribution to the estimate of the probability of failure. At the onset, each patient has

$$\text{potential contribution} = \frac{1}{n},$$

where $n = $ total # of patients under study. Each time a patient fails, the estimate is increased by the amount of contribution of the failed patient. Equivalently, the estimate of survival is decreased by this amount. At a censored time, no death occurred. Hence, the estimate of failure or survival is unchanged at that point in time; that is, the curve neither steps up nor down.

Now, patients who are censored due to lack of follow-up through a specified time still remain capable of failure by this specified time. The potential contribution of these patients cannot be discounted.

> The redistribute-to-the-right algorithm considers a censored patient's potential contribution as being equally redistributed among all patients at risk of failure *after* the censored time.

As a result of this redistribution, any failure that takes place after the censored time contributes slightly more to the estimate of failure than do failures prior to the censored time. That is, the potential contribution of each patient to the estimate increases after the occurrence of a censored patient. Equivalently, K-M, the estimate of survival, decreases (steps down) by this increased equal potential contribution of failures that occur after the censored time. Equivalently, the estimate of survival beyond an observed death time $y_{(i)}$ equals the most recent computed potential contribution times the number of data values to the right of $y_{(i)}$. Table 2.2 implements the redistribute-to-the-right algorithm.

Example with AML1 data:

Let $n_i^* = \#$ at risk just <u>after</u> time $y_{(i)}$, $d_i = \#$ of deaths at time $y_{(i)}$, and $c_i = \#$ of censored observations at time $y_{(i)}$.

Table 2.2: *Redistribute-to-the-right algorithm on AML1 data*

$y_{(i)}$	n_i^*	d_i	c_i	Individual's potential mass at $y_{(i)}$	$\widehat{S}(t)$
0	11	0	0	$\frac{1}{11}$	1
9	10	1	0	$\frac{1}{11}$	$\frac{10}{11} = .91$
13	9	1	0	$\frac{1}{11}$	$\frac{9}{11} = .82$
13	8	0	1	No patient failed; no decrease in K-M. Redistribute equally to the right $\frac{1}{11}$. Each of the 8 remaining patients now has potential mass $$\frac{1}{11} + \left(\frac{1}{11} \times \frac{1}{8}\right) = \frac{1}{11} \times \frac{9}{8}$$	
18	7	1	0	$\frac{1}{11} \times \frac{9}{8}$	$\frac{9}{11} \times \frac{7}{8} = .72$
23	6	1	0	$\frac{1}{11} \times \frac{9}{8}$	$\frac{9}{11} \times \frac{6}{8} = .61$
28	5	0	1	Redistribute $\frac{1}{11} \times \frac{9}{8}$ equally to the right. Each of the 5 remaining patients now has potential mass $$\left(\frac{1}{11} \times \frac{9}{8}\right) + \left(\frac{1}{11} \times \frac{9}{8} \times \frac{1}{5}\right)$$ $$= \frac{1}{11} \times \frac{9}{8} \times \frac{6}{5}$$	
31	4	1	0	$\frac{1}{11} \times \frac{9}{8} \times \frac{6}{5}$	$\frac{9}{11} \times \frac{3}{5} = .49$
34	3	1	0	$\frac{1}{11} \times \frac{9}{8} \times \frac{6}{5}$	$\frac{9}{11} \times \frac{9}{20} = .37$
45	2	0	1	Redistribute $\frac{1}{11} \times \frac{9}{8} \times \frac{6}{5}$ equally to the right. Each of the 2 remaining patients now has potential mass $$\left(\frac{1}{11} \times \frac{9}{8} \times \frac{6}{5}\right) + \left(\frac{1}{11} \times \frac{9}{8} \times \frac{6}{5} \times \frac{1}{2}\right)$$ $$= \frac{1}{11} \times \frac{9}{8} \times \frac{6}{5} \times \frac{3}{2}$$	
48	1	1	0	$\frac{1}{11} \times \frac{9}{8} \times \frac{6}{5} \times \frac{3}{2}$	$\frac{81}{440} = .18$
161	0	0	1		

We revisit this algorithm in the context of competing risks discussed in Chapter 7.2 and of regression quantiles discussed in Chapter 8. With some modification this idea permits generalization to the more general regression quantile models for survival time T. In Chapter 8 we introduce these models and present examples of survival analyses using these models.

Estimate of variance of $\widehat{S}(t)$:

Greenwood's formula (1926):

$$\widehat{\text{var}}\left(\widehat{S}(t)\right) = \widehat{S}^2(t) \sum_{y_{(i)} \leq t} \frac{d_i}{n_i(n_i - d_i)} = \widehat{S}^2(t) \sum_{i=1}^{k} \frac{d_i}{n_i(n_i - d_i)}, \qquad (2.3)$$

where $y_{(k)} \leq t < y_{(k+1)}$.

Example with the AML1 data:

$$\widehat{\text{var}}\left(\widehat{S}(13)\right) = (.82)^2 \left(\frac{1}{11(11-1)} + \frac{1}{10(10-1)}\right) = .0136$$

$$\text{s.e.}\left(\widehat{S}(13)\right) = .1166 \; = \sqrt{.0136}$$

The theory tells us that for each fixed value t

$$\widehat{S}(t) \overset{a}{\sim} \text{normal}\left(S(t), \widehat{\text{var}}\left(\widehat{S}(t)\right)\right).$$

Thus, at time t, an approximate $(1 - \alpha) \times 100\%$ confidence interval for the probability of survival, $S(t) = P(T > t)$, is given by

$$\widehat{S}(t) \pm z_{\frac{\alpha}{2}} \times \text{s.e.}\left(\widehat{S}(t)\right), \qquad (2.4)$$

where s.e. $\left(\widehat{S}(t)\right)$ is the square root of Greenwood's formula for the estimated variance.

Smith (2002), among many authors, discusses the following estimates of hazard and cumulative hazard. Let t_i denote a distinct ordered death time, $i = 1, \ldots, r \leq n$.

Estimates of hazard (risk):

1. Estimate at an observed death time t_i:

$$\widetilde{h}(t_i) = \frac{d_i}{n_i}. \qquad (2.5)$$

2. Estimate of hazard in the interval $t_i \leq t < t_{i+1}$:

$$\widehat{h}(t) = \frac{d_i}{n_i(t_{i+1} - t_i)}. \qquad (2.6)$$

This is referred to as the K-M type estimate. It estimates the rate of death per unit time in the interval $[t_i, t_{i+1})$.

3. Examples with the AML1 data:

$$\tilde{h}(23) = \frac{1}{7} = .143$$

$$\hat{h}(26) = \hat{h}(23) = \frac{1}{7 \cdot (31 - 23)} = .018$$

Estimates of $H(\cdot)$ cumulative hazard to time t:

1. Constructed with K-M:

$$\hat{H}(t) = -\log \hat{S}(t) = -\log \prod_{y_{(i)} \leq t} \left(\frac{n_i - d_i}{n_i} \right), \qquad (2.7)$$

$$\widehat{\operatorname{var}} \left(\hat{H}(t) \right) = \sum_{y_{(i)} \leq t} \frac{d_i}{n_i(n_i - d_i)} . \qquad (2.8)$$

2. Nelson-Aalen estimate (1972, 1978):

$$\tilde{H}(t) = \sum_{y_{(i)} \leq t} \frac{d_i}{n_i}, \qquad (2.9)$$

$$\widehat{\operatorname{var}} \left(\tilde{H}(t) \right) = \sum_{y_{(i)} \leq t} \frac{d_i}{n_i^2} . \qquad (2.10)$$

The Nelson-Aalen estimate is the cumulative sum of estimated conditional probabilities of death from I_1 through I_k where $t_k \leq t < t_{k+1}$. This estimate is the first order Taylor approximation to the first estimate. To see this let $x = d_i/n_i$ and expand $\log(1 - x)$ about $x = 0$.

3. Examples with the AML1 data:

$$\hat{H}(26) = -\log \hat{S}(26) = -\log(.614) = .488$$

$$\tilde{H}(26) = \frac{1}{11} + \frac{1}{10} + \frac{1}{8} + \frac{1}{7} = .4588$$

Estimate of quantiles:

Recall the definition:

the pth-**quantile** t_p is such that $F(t_p) = p$ or $S(t_p) = 1 - p$. As usual, when S is continuous, $t_p \leq S^{-1}(1 - p)$.

As the K-M curve is a step function, the inverse is not uniquely defined. We define the estimated quantile to be

$$\hat{t}_p = \min\{t_i : \hat{S}(t_i) \leq 1 - p\}. \qquad (2.11)$$

By applying the delta method (Chapter 3.2, page 66) to $\widehat{\text{var}}\left(\widehat{S}(\widehat{t}_p)\right)$, Collett (1994, pages 33 and 34) provides the following estimate of variance of $\widehat{t}_p$:

$$\widehat{\text{var}}(\widehat{t}_p) = \frac{\widehat{\text{var}}\left(\widehat{S}(\widehat{t}_p)\right)}{\left(\widehat{f}(\widehat{t}_p)\right)^2},\qquad(2.12)$$

where $\widehat{\text{var}}\left(\widehat{S}(\widehat{t}_p)\right)$ is Greenwood's formula for the estimate of the variance of the K-M estimator, and $\widehat{f}(\widehat{t}_p)$ is the estimated probability density at $\widehat{t}_p$. It is defined as follows:

$$\widehat{f}(\widehat{t}_p) = \frac{\widehat{S}(\widehat{u}_p) - \widehat{S}(\widehat{l}_p)}{\widehat{l}_p - \widehat{u}_p},\qquad(2.13)$$

where $\widehat{u}_p = \max\{t_i | \widehat{S}(t_i) \geq 1 - p + \epsilon\}$, and $\widehat{l}_p = \min\{t_i | \widehat{S}(t_i) \leq 1 - p - \epsilon\}$, for $i = 1, \ldots, r \leq n$ with r being the number of distinct death times, and ϵ a small value. An $\epsilon = 0.05$ would be satisfactory in general, but a larger value of ϵ will be needed if $\widehat{u}_p$ and $\widehat{l}_p$ turn out to be equal. In the following example, we take $\epsilon = 0.05$.

Example with the AML1 data:

The median $\widehat{t}_{.5} = 31$ weeks. We find $\widehat{u}_{.5} = \max\{t_i | \widehat{S}(t_i) \geq 0.55\} = 23$, $\widehat{l}_{.5} = \min\{t_i | \widehat{S}(t_i) \leq 0.45\} = 34$, and $\widehat{f}(31) = \frac{\widehat{S}(23) - \widehat{S}(34)}{34 - 23} = \frac{0.614 - 0.368}{11} = 0.0224$. Therefore, its variance and s.e. are

$$\widehat{\text{var}}(31) = \left(\frac{.1642}{.0224}\right)^2 = 53.73 \quad \text{and} \quad \text{s.e.}(31) = 7.33.$$

An approximate 95% C.I. for the median is given by

$$31 \pm 1.96 \times 7.33 \quad \Rightarrow \quad \left(16.6 \text{ to } 45.4\right) \text{ weeks.}$$

The truncated mean survival time:

The estimated mean is taken to be

$$\widehat{\text{mean}} = \int_0^{y_{(n)}} \widehat{S}(t)\, dt,\qquad(2.14)$$

where $y_{(n)} = \max(y_i)$. If $y_{(n)}$ is uncensored, then this truncated integral is the same as the integral over $[0, \infty)$ since over $[y_{(n)}, \infty)$, $\widehat{S}(t) = 0$. But if the maximum data value is censored, the $\lim_{t \to \infty} \widehat{S}(t) \neq 0$. Thus, the integral over $[0, \infty)$ is undefined. That is, $\widehat{\text{mean}} = \infty$. To avoid this we truncate the integral. By taking the upper limit of integration to be the $y_{(n)}$, we redefined the K-M estimate to be zero beyond the largest observation. Another way to look at this is that we have forced the largest observed time to be uncensored. This does give, however, an estimate biased towards zero. This estimate is the total area under the K-M curve. As $\widehat{S}(t)$ is a step function, we compute this area as the following sum:

$$\widehat{\text{mean}} = \sum_{i=1}^{n'} \left(y_{(i)} - y_{(i-1)} \right) \widehat{S}(y_{(i-1)}), \tag{2.15}$$

where $n' = \#$ of distinct observed y_i's, $n' \leq n$, $y_{(0)} = 0$, $\widehat{S}(y_{(0)}) = 1$, and $\widehat{S}(y_{(i-1)})$ is the height of the function at $y_{(i-1)}$.

In the AML1 data, $y_{(n)} = 161$ and, from the following S output, the estimated expected survival time $\widehat{\text{mean}} = 52.6$ weeks. The variance formula for this estimator is given in Remark 5. An estimate of the truncated mean residual life, mrl(t), along with a variance estimate is given in Remark 6.

Note: As survival data are right skewed, the median is the preferred descriptive measure of the typical survival time.

`survfit`:

This is the main S nonparametric survival analysis function. Its main argument takes a `Surv(time,status)` object. We have modified some of the output. Data for both groups in the AML study are in a data frame called **aml**. The "group" variable $= 1$ for maintained group, $= 0$ for nonmaintained.

```
> aml1 <- aml[aml$group == 1, ] # Creates a data frame with
                                # maintained group data only.
> Surv(aml1$weeks,aml1$status) # Surv object
[1]   9    13    13+   18    23    28+   31    34    45+   48    161+
> km.fit <- survfit(Surv(weeks,status),type="kaplan-meier",
            data = aml1)
> plot(km.fit,conf.int=F,xlab="time until relapse (in weeks)",
       ylab="proportion in remission",lab=c(10, 10, 7))
> mtext("K-M survival curve for the AML data",3,line=-1,cex=2)
> mtext("maintained group",3,line = -3)
> abline(h=0) # Figure 2.3 is now complete.
> km.fit
      n events mean se(mean) median 0.95LCL 0.95UCL
     11      7 52.6     19.8     31      18      NA
> summary(km.fit) # survival is the estimated S(t).
 time n.risk n.event survival std.err    95% LCL    95% UCL
    9     11       1    0.909  0.0867     0.7541      1.000
   13     10       1    0.818  0.1163     0.6192      1.000
   18      8       1    0.716  0.1397     0.4884      1.000
   23      7       1    0.614  0.1526     0.3769      0.999
   31      5       1    0.491  0.1642     0.2549      0.946
```

```
   34      4      1     0.368  0.1627        0.1549        0.875
   48      2      1     0.184  0.1535        0.0359        0.944
> attributes(km.fit) # Displays the names of objects we can
                     # access.
$names:
[1] "time" "n.risk" "n.event" "surv" "std.err" "upper"
[7] "lower" "conf.type" "conf.int" "call"
$class: [1] "survfit"
# Example: to access "time" and "surv"
> t.u <- summary(km.fit)$time # t.u is a vector with the
                             # seven uncensored times.
> surv.u <- summary(km.fit)$surv # Contains the estimated
                                # S(t.u).
```

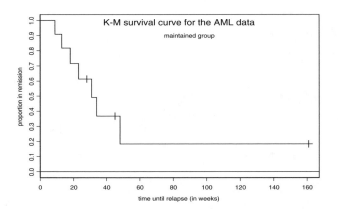

Figure 2.3 *Kaplan-Meier survival curve. A + indicates a censored value.*

Remarks:

1. Notice the effect of accommodating the censored data points. The median time in complete remission is increased from 28 weeks to 31 weeks. The expected time is increased from 38.45 weeks to 52.6 weeks. This explains the third method alluded to in the **A naive descriptive analysis of AML study** presented in Chapter 1.1, page 4.

2. survfit uses a simple graphical method of finding a confidence interval for the median. Upper and lower confidence limits for the median are defined in terms of the confidence intervals for $S(t)$: the upper confidence limit is the smallest time at which the upper confidence limit for $S(t)$ is ≤ 0.5. Likewise, the lower confidence limit is the smallest time at which the lower confidence limit for $S(t)$ is ≤ 0.5. That is, draw a horizontal line at 0.5 on

the graph of the survival curve, and use intersections of this line with the curve and its upper and lower confidence bands. If, for example, the UCL for $S(t)$ never reaches 0.5, then the corresponding confidence limit for the median is unknown and it is represented as an NA. See pages 242 and 243, S-PLUS 2000, Guide to Statistics, Vol.II.

3. Confidence intervals for pth-quantile without using an estimate of the density (2.13) at $\hat{t}_p$ are also available. See Chapter 4.5, Klein & Moeschberger (1997).

4. The default confidence intervals for $S(t)$ produced by survfit are **not** constructed solely with the Greenwood's standard errors (std.err) provided in the output. To obtain confidence intervals which use the Greenwood's s.e. directly, you must specify conf.type="plain" in the survfit function. These correspond to the formula (2.4).

The **default** intervals in survfit are called "log" and the formula is:

$$\exp\left(\log \hat{S}(t) \pm 1.96\,\text{s.e.}\left(\hat{H}(t)\right)\right), \qquad (2.16)$$

where $\hat{H}(t)$ is the estimated cumulative hazard function (2.7) with s.e. being $\approx$ the square root of the variance (2.8). These "log" intervals are derived using the delta method defined in Chapter 3.2, page 66. The log-transform on $\hat{S}(t)$ produces more efficient intervals as we remove the source of variation due to using $\hat{S}(t)$ in the variance estimate. Hence, this approach is preferred.

Sometimes, both of these intervals give limits outside the interval $[0, 1]$. This is not so appealing as $S(t)$ is a probability! Kalbfleisch & Prentice (1980) suggest using the transformation $W = \log(-\log(\hat{S}(t)))$ to estimate the log cumulative hazard parameter $\log(-\log(S(t)))$, and to then transform back. Using the delta method, an estimate of the asymptotic variance of this estimator is given by

$$\widehat{\text{var}}(W) \approx \frac{1}{(\log \hat{S}(t))^2}\widehat{\text{var}}(-\log \hat{S}(t)) = \frac{1}{(\log \hat{S}(t))^2}\sum_{y_{(i)} \leq t}\frac{d_i}{n_i(n_i - d_i)}. \qquad (2.17)$$

An approximate $(1 - \alpha) \times 100\%$ C.I. for the quantity $S(t)$ is given by

$$\left(\hat{S}(t)\right)^{\exp\{z_{\frac{\alpha}{2}}\,\text{s.e.}(W)\}} \leq S(t) \leq \left(\hat{S}(t)\right)^{\exp\{-z_{\frac{\alpha}{2}}\,\text{s.e.}(W)\}}. \qquad (2.18)$$

To get these intervals specify conf.type="log-log" in the survfit function. These intervals will always have limits within the interval $[0, 1]$.

5. The variance of the estimated truncated mean survival time (2.14) is

$$\widehat{\text{var}}(\widehat{\text{mean}}) = \sum_{i=1}^{n'}\left(\int_{y_{(i)}}^{y_{(n)}}\hat{S}(u)du\right)^2\frac{d_i}{n_i(n_i - d_i)}. \qquad (2.19)$$

The quantity se(mean) reported in the survfit output is the square root of this estimated variance.

6. An estimate of the truncated mean residual life at time t (1.9), denoted by $\widehat{\mathrm{mrl}}(t)$, is taken to be

$$\widehat{\mathrm{mrl}}(t) = \frac{\int_t^{y_{(n)}} \hat{S}(u)du}{\widehat{S}(t)} \tag{2.20}$$

with estimated variance

$$\widehat{\mathrm{var}}\left(\widehat{\mathrm{mrl}}(t)\right) = \frac{1}{\widehat{S}^2(t)} \left(\sum_{t \leq y_{(i)} \leq y_{(n)}} \left(\int_{y_{(i)}}^{y_{(n)}} \widehat{S}(u)du \right)^2 \frac{d_i}{n_i(n_i - d_i)} \right.$$
$$\left. - \not\!\!\frac{2}{}\left(\int_t^{y_{(n)}} \widehat{S}(u)du \right)^2 \sum_{y_{(i)} \leq t} \frac{d_i}{n_i(n_i - d_i)} \right). \tag{2.21}$$

To derive this estimate of variance one needs to use the bivariate delta method in Chapter 3.6 as $\widehat{\mathrm{mrl}}(t)$ is a quotient of two estimators.

7. This remark is for the more advanced reader who is interested in large sample theory. Breslow and Crowley (1974) proved, for t in the interval $[0, A]$ with $F(A) < 1$,

$$Z_n(t) = \sqrt{n}\left(\widehat{S}(t) - S(t)\right) \text{ converges weakly to } Z(t)$$
$$\text{where } Z(t) \text{ is a Gaussian process.}$$

Hence, $Z(t_1), \ldots, Z(t_k)$ has a multivariate normal distribution for $t_1, \ldots, t_k$ and this distribution is completely determined by the marginal means and the covariance values. As a consequence, $\widehat{S}(t)$ has asymptotic mean $S(t)$ and asymptotic covariance structure

$$\frac{1}{n} S_f(t_1) S_f(t_2) \times \int_0^{\min(t_1, t_2)} \frac{|dS_f(u)|}{S_f(u)S_m(u)},$$

where $S_f = 1 - F$ and S_m is the survivor function of $Y = \min(T, C)$ discussed in Chapter 1.3, page 13. Further, for any function g continuous in the sup norm, $g(Z_n(t))$ converges in distribution to $g(Z(t))$. With these facts established, the derivation of the variance formulae in Remarks 5 and 6 are easily verified. We assume expectation E can be passed through the integral $\int_a^b$. For a gentle introduction to stochastic integrals (which the estimators of the mean and mrl are) and Gaussian processes, see Ross (2000).

The hazard.km and quantile.km functions:

The function hazard.km takes a survfit object for its argument. It outputs $\hat{h}(t)$, $\tilde{h}(t_i)$, $\widehat{H}(t)$, se($\widehat{H}(t)$), $\widetilde{H}(t)$, and se($\widetilde{H}(t)$). The function quantile.km computes an estimated pth-quantile along with its standard error and an approximate $(1 - \alpha) \times 100\%$ confidence interval. It has four arguments: (data,p,eps,z), where data is a survfit object, p is a scalar between 0 and 1, eps (ϵ) is .05 or a little larger, and z is the standard normal z-score needed for the desired confidence level.

```
> hazard.km(km.fit)         T(t.)   H(t)        H(t)
     time ni di  hihat hitilde   Hhat  se.Hhat Htilde se.Htilde
  1     9 11   1 0.0227  0.0909 0.0953  0.0953 0.0909    0.0909
  2    13 10   1 0.0200  0.1000 0.2007  0.1421 0.1909    0.1351
  3    18  8   1 0.0250  0.1250 0.3342  0.1951 0.3159    0.1841
  4    23  7   1 0.0179  0.1429 0.4884  0.2487 0.4588    0.2330
  5    31  5   1 0.0667  0.2000 0.7115  0.3345 0.6588    0.3071
  6    34  4   1 0.0179  0.2500 0.9992  0.4418 0.9088    0.3960
  7    48  2   1     NA  0.5000 1.6923  0.8338 1.4088    0.6378
> quantile.km(km.fit,.25,.05,1.96) # the .25th-quantile
[1] "summary"
   qp se.S.qp  f.qp  se.qp   LCL    UCL
1 18   0.1397 0.0205 6.8281 4.617 31.383  # in weeks
```

Remarks:

1. In the case of no censoring, quantile.km differs from the S function quantile. Try quantile(1:10,c(.25,.5,.75)) and compare quantile.km after using survfit(Surv(1:10,rep(1,10))).

2. If we extend the survfit graphical method to find the confidence limits for a median to the .25th quantile, we get 13 and NA as the lower and upper limits, respectively. WHY! See Remark 2, page 36.

2.2 Comparison of survivor curves: two-sample problem

For the AML data the variable "weeks" contains all 23 observations from both groups.
There is now the variable group:

$$\text{group} = \begin{cases} 1 & \text{for maintained} \\ 0 & \text{for nonmaintained.} \end{cases}$$

A plot of the K-M curves for both groups is displayed in Figure 2.4. A summary of the survival estimation using the survfit function follows:

```
> km.fit <- survfit(Surv(weeks,status)~group,data=aml)
> plot(km.fit,conf.int=F,xlab="time until relapse (in weeks)",
           ylab="proportion without relapse",
              lab=c(10,10,7),cex=2,lty=1:2)
```

```
> summary(km.fit) # This displays the survival probability
                  # table for each group. The output is omitted.
> km.fit
            n  events  mean  se(mean)  median 0.95LCL 0.95UCL
group=0 12      11    22.7     4.18       23       8      NA
group=1 11       7    52.6    19.83       31      18      NA
```

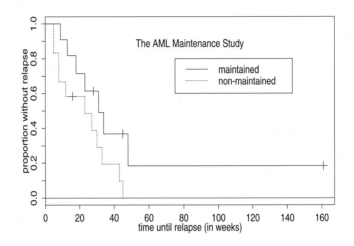

Figure 2.4 *A comparison of two K-M curves.*

• Notice the estimated mean, median, and survivor curve of "maintained" group are higher than those of the other group.

• Is there a significant difference between the two survivor curves? Does maintenance chemotherapy statistically prolong time until relapse?

To test $H_0 : F_1 = F_2$, we present the Mantel-Haenszel (1959) test, also called the **log-rank test**. Another well known test is the Gehan (1965) test, which is an extension of the Wilcoxon test to accommodate right-censored data. See Miller (1981, Chapter 4.1) for a presentation of this test. To motivate the construction of the Mantel-Haenszel test statistic, we first briefly study Fisher's exact test.

Comparing two binomial populations:

Suppose we have two populations, and an individual in either population can have one of two characteristics. For example, Population 1 might be cancer patients under a certain treatment and Population 2 cancer patients under a different treatment. The patients in either group may either die within a year or survive beyond a year. The data are summarized in a 2×2 contingency table. Our interest here is to compare the two binomial populations, which is common in medical studies.

	Dead	Alive	
Population 1	a	b	n_1
Population 2	c	d	n_2
	m_1	m_2	n

Denote

$$\begin{aligned} p_1 &= P\{\text{Dead}|\text{Population 1}\}, \\ p_2 &= P\{\text{Dead}|\text{Population 2}\}. \end{aligned}$$

Want to test

$$H_0 : p_1 = p_2.$$

Fisher's exact test:

The random variable A, which is the entry in the $(1,1)$ cell of the 2×2 table, has the following **exact discrete conditional distribution under $\mathbf{H_0}$**: Given n_1, n_2, m_1, m_2 fixed quantities, it has a **hypergeometric distribution** where

$$P\{A = a\} = \frac{\binom{n_1}{a}\binom{n_2}{m_1-a}}{\binom{n}{m_1}}.$$

The test based on this exact distribution is called the **Fisher's exact test**. The S function `fisher.test` computes an exact p-value. The mean and variance of the hypergeometric distribution are

$$\begin{aligned} E_0(A) &= \frac{n_1 m_1}{n}, \\ Var_0(A) &= \frac{n_1 n_2 m_1 m_2}{n^2(n-1)}. \end{aligned}$$

We can also conduct an approximate chi-square test when samples are large as

$$\chi^2 = \left(\frac{a - E_0(A)}{\sqrt{Var_0(A)}}\right)^2 \overset{a}{\sim} \chi_{(1)},$$

where $\chi_{(1)}$ denotes a chi-square random variable with 1 degree of freedom.

Mantel-Haenszel/log-rank test:

Now suppose we have a sequence of 2×2 tables. For example, we might have k hospitals; at each hospital, patients receive either Treatment 1 or Treatment 2 and their responses are observed. Because there may be differences among hospitals, we do not want to combine all k tables into a single 2×2 table. We want to test

$$H_0 : p_{11} = p_{12}, \text{ and} \ldots, \text{ and } p_{k1} = p_{k2},$$

where

$$p_{i1} = P\{\text{Dead}|\text{Treatment } 1, \text{Hospital } i\},$$
$$p_{i2} = P\{\text{Dead}|\text{Treatment } 2, \text{Hospital } i\}.$$

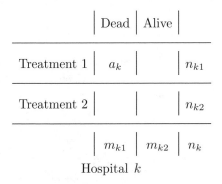

	Dead	Alive	
Treatment 1	a_1		n_{11}
Treatment 2			n_{12}
	m_{11}	m_{12}	n_1

Hospital 1

$\vdots$

	Dead	Alive	
Treatment 1	a_k		n_{k1}
Treatment 2			n_{k2}
	m_{k1}	m_{k2}	n_k

Hospital k

Use the **Mantel-Haenszel (1959) statistic**

$$\text{MH} = \frac{\sum_{i=1}^{k}(a_i - E_0(A_i))}{\sqrt{\sum_{i=1}^{k} Var_0(A_i)}}. \tag{2.22}$$

If the tables are independent, then MH $\overset{a}{\sim}$ $N(0, 1)$ either when k is fixed and $n_i \to \infty$ or when $k \to \infty$ and the tables are also identically distributed.

In survival analysis the MH statistic is applied as follows: Combine the two samples, order them, and call them $z_{(i)}$. Construct a 2×2 table for each uncensored time point $z_{(i)}$. Compute the MH statistic for this sequence of tables to test $H_0 : F_1 = F_2$. The theory tells us that asymptotic normality still holds even though these tables are clearly not independent.

We illustrate how to compute the MH with the following fictitious data:

Treatment Old	3, 5, 7, 9+, 18
Treatment New	12, 19, 20, 20+, 33+

Computations for the MH are given in the following table. Denote the combined ordered values by z. Note that n is the total number of patients at risk in both groups; m_1 the number of patients who died at the point z; n_1 the number at risk in treatment Old at time z; a equals 1 if death in Old or 0 if death in New. Remember that

$$E_0(A) = \frac{m_1 n_1}{n} \quad \text{and} \quad Var_0(A) = \frac{m_1(n - m_1)}{n - 1} \times \frac{n_1}{n}\left(1 - \frac{n_1}{n}\right).$$

trt	z	n	m_1	n_1	a	$E_0(A)$	r	$\frac{m_1(n-m_1)}{n-1}$	$\frac{n_1}{n}\left(1 - \frac{n_1}{n}\right)$
Old	3	10	1	5	1	.50	.50	1	.2500
Old	5	9	1	4	1	.44	.56	1	.2469
Old	7	8	1	3	1	.38	.62	1	.2344
Old	9+		0		0				
New	12	6	1	1	0	.17	−.17	1	.1389
Old	18	5	1	1	1	.20	.80	1	.1600
New	19	4	1	0	0	0	0	1	0
New	20	3	1	0	0	0	0	1	0
New	20+								
New	33+								
Total					4	1.69	2.31		1.0302

where $r = (a - E_0(A))$. Then

$$\text{MH} = \frac{\text{sum of } (a - E_0(A))}{\sqrt{\text{sum of }\left(\frac{m_1(n-m_1)}{n-1} \times \frac{n_1}{n}\left(1 - \frac{n_1}{n}\right)\right)}}$$

$$= \frac{2.31}{1.02} = 2.26$$

$$p\text{-value} = 0.012 \quad (\text{one-tailed } Z \text{ test}).$$

The S function `survdiff` provides the log-rank (= MH) test by default. Its first argument takes a `Surv` object. It gives the square of the MH statistic which is then an approximate chi-square statistic with 1 degree of freedom. This is a two-tailed test. Hence, the p-value is twice that of the MH above. Except for round-off error, everything matches.

```
> grouph <- c(1,1,1,1,1,2,2,2,2,2) # groups: 1=old; 2=new
> hypdata <- c(3,5,7,9,18,12,19,20,20,33) # the data
> cen <- c(1,1,1,0,1,1,1,1,0,0) # censor status:
                                # 1=uncensored; 0=censored
> survdiff(Surv(hypdata,cen)~grouph)
```

	N	Observed	Expected	(O-E)^2/E	(O-E)^2/V
grouph=1	5	4	1.69	3.18	5.2
grouph=2	5	3	5.31	1.01	5.2

```
Chisq = 5.2  on 1 degrees of freedom, p = 0.0226
# This p-value corresponds to a two-tailed Z-test
# conducted with MH.
> sqrt(5.2) # square root of log-rank test statistic.
[1] 2.280351 # MH.
# .0226 = (1 - pnorm(2.280351))*2: p-value for two-sided test
> .0226/2
[1] 0.0113 # p-value for one-sided test.
```

The log-rank test on the AML data is:

```
> survdiff(Surv(week,status)~group,data=aml)
```

	N	Observed	Expected	(O-E)^2/E	(O-E)^2/V
group=1	11	7	10.69	1.27	3.4
group=2	12	11	7.31	1.86	3.4

```
Chisq= 3.4  on 1 degrees of freedom, p= 0.0653
```

There is mild evidence to suggest that maintenance chemotherapy prolongs the remission period since the one-sided test is appropriate and its p-value is $.0653/2 = .033$.

Remarks:

1. The proofs of asymptotic normality of the K-M and MH statistics are not easy. They involve familiarity with Gaussian processes and the notion of weak convergence in addition to other probability results applied to stochastic processes. Here the stochastic processes are referred to as empirical processes. For the interested reader, these proofs of normality are outlined in Miller (1981), Chapters 3.2.5, 3.3, 4.2.4. For the even more advanced reader, see Fleming and Harrington (1991).

2. One might ask why the MH test is also called the log-rank test. One explanation is to consider the accelerated failure time model presented in Chapter 4.4. Suppose there is only one covariate x so that the model is $Y = \beta_0^* + \beta^* x + \sigma Z$. Miller (1981), Chapter 6.2.1, derives the locally most powerful rank statistic for testing $H_0 : \beta^* = 0$ against $H_A : \beta^* \neq 0$ when evaluated at the extreme value distribution for error. He says that Peto and Peto (1972) first derived this test and named it the **log-rank test**. When x is binary, that is, $x = 1$ if in group 1 and $= 0$ if in group 2, this statistic simplifies to a quantity that is precisely a rescaled version of the MH statistic when there are no ties. The locally most powerful rank statistic involves a log, hence the name log-rank test.

3. The `survdiff` function contains a "rho" parameter. The default value, rho $= 0$, gives the log-rank test. When rho $= 1$, this gives the Peto test. This test was suggested as an alternative to the log-rank test by Prentice and Marek (1979). The Peto test emphasizes the beginning of the survival curve in that earlier failures receive larger weights. The log-rank test emphasizes the tail of the survival curve in that it gives equal weight to each failure time. Thus, choose between the two according to the interests of the study. The choice of emphasizing earlier failure times may rest on clinical features of one's study.

Hazard ratio as a measure of effect:

The hazard ratio is a descriptive measure of the treatment (group) effect on survival. Here we use the two types of empirical hazard functions, $\tilde{h}$ and $\hat{h}(t)$, defined on page 32, to form ratios and then interpret them in the context of the AML study. The function `emphazplot` contains an abridged form of the `hazard.km` function (page 38) and produces two plots, one for each of the two types of hazard estimates. Modified output and plots follow.

```
> attach(aml)
> Surv0 <- Surv(weeks[group==0],status[group==0])
> Surv1 <- Surv(weeks[group==1],status[group==1])
> data <- list(Surv0,Surv1)
> emphazplot(data,text="solid line is maintained group")
```

	nonmaintained				maintained		
	time	hitilde	hihat		time	hitilde	hihat
1	5	0.167	0.056	1	9	0.091	0.023
2	8	0.200	0.050	2	13	0.100	0.020
3	12	0.125	0.011	3	18	0.125	0.025
4	23	0.167	0.042	4	23	0.143	0.018
5	27	0.200	0.067	5	31	0.200	0.067
6	30	0.250	0.083	6	34	0.250	0.018
7	33	0.333	0.033	7	48	0.500	0.018
8	43	0.500	0.250				
9	45	1.000	0.250				

> detach()

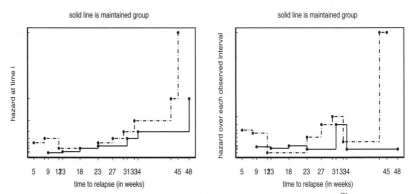

Figure 2.5 *A comparison of empirical hazards. Left plot displays* $\widetilde{h}(t_i)$. *Right plot displays* $\widehat{h}(t)$.

Consider the following two hazard ratios of nonmaintained to maintained:

$$\frac{\widehat{h}_{\mathrm{nm}}(15)}{\widehat{h}_{\mathrm{m}}(15)} = \frac{.011}{.020} = .55 \qquad \text{and} \qquad \frac{\widehat{h}_{\mathrm{nm}}(25)}{\widehat{h}_{\mathrm{m}}(25)} = \frac{.042}{.018} = 2.33 \; .$$

The nonmaintained group has 55% of the risk of those maintained of relapsing at 15 weeks. However, on the average, those nonmaintained have 2.33 times the risk of those maintained of relapsing at 25 weeks.

Neither of the two plots in Figure 2.5 displays roughly parallel curves over time. In the second plot, the hazard curves cross over time, which implies one group's risk is not always lower than the other's with respect to time. **Both plots indicate the hazard ratio is not constant with respect to time**, which says the hazard functions of the two groups are not proportional. The notion of proportional hazards is a central theme threaded throughout survival analyses. It is discussed in detail in Chapters 4, 5, 6, 7, and 8. Note these plots in Figure 2.5 are only an illustration of how to visualize and interpret HR's. Of course, statistical accuracy (confidence bands) should be incorporated as these comments may not be statistically significant.

Stratifying on a covariate:

- Stratifying on a particular covariate is one method that can account for (adjust for) its possible confounding and/or interaction effects with the treatment of interest on the response.

- Confounding and/or interaction effects of other known factors with the treatment variable **can mask the "true" effects of the treatment of interest.** Thus, stratification can provide us with stronger (or weaker) evidence, or more importantly, reverse the sign of the effect. That is, it is possible for the aggregated data to suggest treatment is favorable when in fact, in every subgroup, it is highly unfavorable; and vice versa. This is known as **Simpson's paradox** (Simpson, 1951).

Let's consider the fictitious data again and see

1. What happens when we stratify by sex?
2. How is the log-rank statistic computed?

Recall:

```
grouph <- c(1,1,1,1,1,2,2,2,2,2) # groups: 1 = old 2 = new
hypdata <- c(3,5,7,9,18,12,19,20,20,33) # the data
cen <- c(1,1,1,0,1,1,1,1,0,0) # censor status:
                              1 = uncensored; 0 = censored
```

How to:

Separate the data by sex. Then, within each sex stratum, construct a sequence of tables as we did above. Then combine over the two sexes to form $(\text{MH})^2$. According to the sex vector

$$\text{sex} <- c(\overbrace{1,1,1,2,2}^{\text{old}},\overbrace{2,2,2,1,1}^{\text{new}}), \quad \text{where } 1 = \text{male} \quad 2 = \text{female.}$$

Within each stratum, n is the total number at risk, m_1 the number who die at point z, n_1 the number at risk in treatment Old at time z, and a equals 1 if death in Old or 0 if death in New.

$$\textbf{MALE}: \quad \begin{array}{ll} \text{Old} & 3,\ 5,\ 7 \\ \text{New} & 20+,\ 33+ \end{array}$$

trt	z	n	m_1	n_1	a	$E_0(A)$	$\frac{m_1(n-m_1)}{n-1}$	$\frac{n_1}{n}\left(1-\frac{n_1}{n}\right)$
Old	3	5	1	3	1	.60	1	.24
Old	5	4	1	2	1	.50	1	.25
Old	7	3	1	1	1	.333333	1	.222222
New	20+	2						
New	33+	1						
Total					3	1.433333		.712222

Note: $E_0(A) = \frac{n_1 m_1}{n}$ and $Var_0(A) = \frac{m_1(n-m_1)}{n-1} \times \frac{n_1}{n}\left(1-\frac{n_1}{n}\right)$.

FEMALE :   Old   9+, 18
                New   12, 19, 20

trt	z	n	m_1	n_1	a	$E_0(A)$	$\frac{m_1(n-m_1)}{n-1}$	$\frac{n_1}{n}\left(1-\frac{n_1}{n}\right)$
Old	9+	5						
New	12	4	1	1	0	.25	1	.1875
Old	18	3	1	1	1	.333333	1	.222222
New	19	2	1	0	0	0		0
New	20	1	1	0	0	0		0
Total					1	.583333		.409722

Then pooling by summing over the two tables, we have $a = 4$, $E_0(A) = 1.433333 + .583333 = 2.016666$, and $Var_0(A) = .712222 + .409722 = 1.121944$. The log-rank statistic is

$$(\text{MH})^2 = \frac{(4 - 2.016666)^2}{1.121944} = 3.506,$$

which matches the following S output from `survdiff`. Note the `strata(sex)` term that has been included in the model statement within the `survdiff` function.

```
# sex = 1 for male, sex = 2 for female
# group = 1 for old, group = 2 for new treatment

> survdiff(Surv(hypdata,cen)~grouph+strata(sex))

          N Observed Expected (O-E)^2/E (O-E)^2/V
grouph=1  5    4       2.02     1.951      3.51
grouph=2  5    3       4.98     0.789      3.51

Chisq= 3.5  on 1 degrees of freedom, p= 0.0611
```

Note that the p-value of a one-sided alternative is $0.0611/2 = .031$. Although there is still significant evidence at the .05 level that the new treatment is better, it is not as strong as before we stratified. That is, after taking into account the variation due to sex, the difference between treatments is not as strong.

The next example shows that stratification can even reverse the association.

Example of Simpson's paradox:

This example is extracted from an article written by Morrell (1999). He discusses data collected in a South African longitudinal study of growth of children, referred to as the Birth to Ten study (BTT).

Extract: This study commenced in the greater Johannesburg/Soweto metropolitan area of South Africa during 1990. A birth cohort was formed from all singleton births during a seven-week period between April and June 1990 to women with permanent addresses within a defined area. Identification of children born during this seven-week period and living in the defined areas took place throughout the first year of the study, by the end of which 4029 births had been enrolled. The BTT study collected prenatal, birth, and early development information on these children. The aim of the study was to identify factors related to the emergence of cardiovascular disease risk factors in children living in an urban environment in South Africa. In 1995, when the children were five years old, the children and care-givers were invited to attend interviews. Detailed questionnaires were completed that included questions about living conditions within the child's home, the child's exposure to tobacco smoke, and additional health-related issues. The five-year sample consisted of 964 children. Unfortunately, there was a great deal of missing data in the baseline group, especially on the variables reported below.

If the five-year sample is to be used to draw conclusions about the entire birth cohort, the five-year group should have characteristics similar to those who were not traced from the initial group. Thus, the five-year group was compared to those who did not participate in the five-year interview on a number of factors. One of the factors was a variable that determined whether the mother had medical aid (which is similar to health insurance) at the time of the birth of the child.

Table 2.3 shows that 11.1% of those in the five-year cohort had medical aid, whereas 16.6% of those who were not traced had medical. This difference is statistically significant (p-value $= 0.007$). The subjects in the BTT study were also classified by their racial group. In this article we consider only white and black participants. Tables 2.4 and 2.5 show the distribution of the medical aid variable broken down by race (two strata). For whites, 83.3% of those in the five-year cohort had medical aid, whereas 82.5% of those who did not participate in the five-year tests had medical aid. For blacks, the corresponding percentages are 8.9% and 8.7%. This shows that even though overall a significantly smaller percentage of the five-year cohort had medical aid, when the race of the subjects is taken into account, the association is reversed. Furthermore, there is negligible evidence of any difference between the percentages when stratified by race; p-value $= 0.945$ and 0.891 for whites and blacks, respectively. The $(MH)^2$ statistic (page 42), which pools effects across the two race tables, has a value of 0.0025 with p-value = 0.9599.

Table 2.3: *Number and percentages of mothers with medical aid*

	Children Not Traced	Five-Year Group	Total
Had Medical Aid	195 (16.61%)	46 (11.06%)	241
No Medical Aid	979 (83.39%)	370 (88.94%)	1349
Total	1174 (100%)	416 (100%)	1590

Table 2.4: *Number and percentages of mothers with medical aid (white)*

	Children Not Traced	Five-Year Group	Total
Had Medical Aid	104 (82.54%)	10 (83.33%)	114
No Medical Aid	22 (17.46%)	2 (16.67%)	24
Total	126 (100%)	12 (100%)	138

Table 2.5: *Number and percentages of mothers with medical aid (black)*

	Children Not Traced	Five-Year Group	Total
Had Medical Aid	91 (8.68%)	36 (8.91%)	127
No Medical Aid	957 (91.32%)	368 (91.09%)	1325
Total	1048 (100%)	404 (100%)	1452

This reversal, and elimination, of association is easily explained. Whites tend to have more access to medical aid than do black South Africans (83% and 8.7%, respectively). In addition, many more blacks were originally included in the BTT study than whites (1452 blacks, 138 whites). Consequently, when the race groups were combined, a relatively small percentage (241/1590 = 15.16%) of the subjects have access to medical aid. At the five-year follow-up, very few whites agree to attend the interviews (12/138 = 8.67% of those with data on the medical aid variable). Possibly whites felt they had little to gain from participating in the study, while a larger proportion of blacks (404/1452 = 27.82% of those with data on the medical aid variable) continue into the five-year study. The blacks may have valued the medical checkup and screening provided to children in the study as a replacement for (or in addition to) a regular medical screening.

The data contained in the above tables are found in the S data frame BirthtoTen and in BirthtoTen.xls.

Key to variables in BirthtoTen data:		
Medical.Aid	0 = No,	1 = Yes
Traced	0 = No,	1 = Five-Year Cohort
Race	1 = White,	2 = Black

The S function `crosstabs` produces contingency tables. The S function `mantelhae.test` conducts the $(MH)^2$ test when pooling 2×2 tables. The following S code produces the foregoing results:

```
> crosstabs(~ Medical.Aid + Traced,data=BirthtoTen)
> crosstabs(~ Medical.Aid + Traced,data=BirthtoTen,
                            subset=Race==1)
> crosstabs(~ Medical.Aid+Traced,data=BirthtoTen,
                            subset=Race==2)
> mantelhaen.test(BirthtoTen$Medical.Aid,BirthtoTen$Traced,
                            BirthtoTen$Race)
```

2.3 Exercises

A. *Applications*

2.1 Use only hand-held calculator. No need for computer.

(a) Calculate the following table and sketch the Kaplan-Meier (K-M) estimate of survival for the data set y: 1, 1^+, 2, 4, 4, 4^+, 6, 9. ("+" denotes censored observation.) The s.e.$(\widehat{S}(t))$ is computed using Greenwood's formula (2.3) for the estimated (asymptotic) variance of the K-M curve at time t.

$y_{(i)}$	d_i	n_i	$\widehat{p}_i$	$\widehat{S}(y_{(i)}) = \widehat{P}(T > y_{(i)})$	s.e.$(\widehat{S}(y_{(i)}))$
0					
1					
1^+					
2					
4					
4^+					
6					
9					

(b) Calculate a 95% confidence interval for $S(t)$ at $t = 3$. Use the **default interval** given in Remark 4, expression (2.16). Is is necessary to use the C.I. in expression (2.18)? If yes, use it to report a 95% C.I.

(c) Compute the estimated hazard (2.5) $\widetilde{h}(t_i)$ at $t_i = 2$. Then compute a 95% C.I. for $H(t)$ at $t = 3$ using the Nelson-Aalen estimate (2.9).

(d) Provide a point and 95% C.I. estimate of the median survival time. See page 34.

2.2 For AML1 data, compute by hand point and 95% C.I. estimates of the .25th quantile using formulae (2.11), (2.12), and (2.13). Check your answers with the results given by the function `quantile.km` on page 39.

2.3 Use S or R for this exercise.

In this study the survival times (in days) of 66 patients after a particular operation were observed. The data frame **diabetes** contains for each patient the following variables:

Variable	Key
sex	gender (m=0, f=1)
diab	diabetic (1=yes, 0=no)
alter	age in years
altgr	age group in years = 0 if age $\leq$ 64, or 1 if age $>$ 64
lzeit	survival times in days (number of days to death) after operation
tod	0 = censored, 1 = uncensored (dead)

(a) Following the S code on page 35 of the text, obtain a summary of the K-M survival curve for the **diabetic group only**. `survfit` is the main function.

(b) Report the mean and median survival times.

(c) Plot the K-M curve for this group.

(d) Use the function `hazard.km` (page 38) to give a summary of the various estimates of hazard and cumulative hazard.

(e) Use the function `quantile.km` (page 38) to provide point and 95% confidence interval estimates for the .25th and .80th quantiles.

2.4 We continue with the diabetes data.

(a) Plot the K-M curves for the data of the diabetic group and the nondiabetic. Comment briefly! Be sure to give a legend so we know which line corresponds to which group. See page 39.

(b) Is there a statistically significant difference in survival between the two groups – diabetic and nondiabetic? What weaknesses (shortcomings) does this global analysis have? See page 46 for example.

(c) Stratifying on known prognostic factor(s) can improve an analysis.

i. Stratify by `sex`. Judge now the difference in survival between the diabetic and nondiabetic groups.

Tips:

```
> table(diabetes$sex,diabetes$diab) # This counts the
# subjects in each gender (stratum) for each group.
> survdiff(Surv(lzeit,tod)~diab+strata(sex),
                                data=diabetes)
> fit.sex <- survfit(Surv(lzeit,tod)~diab+strata(sex))
> fit.sex
> plot(fit.sex,lty=1:4)
```

ii. Stratify by `altgr` (age group). Judge now the difference in survival between the diabetic and nondiabetic groups.

(d) Refer to the **Hazard ratio as a measure of effect** discussion starting on page 45. Does it appear that the hazard ratio between the two groups, diabetic and nondiabetic, is constant over time? That is, are the two empirical hazard functions proportional?

Tip:

Use the function `emphazplot`.

B. *Theory and WHY!*

2.5 Answer the WHY! on page 39.

2.6 Show the K-M estimator (2.2) reduces to the **esf** (2.1) when there are no censored observations.

2.7 On the data given in Exercise 2.1, compute by hand the truncated mean survival time (2.15) and its estimated variance (2.19). Check your answer using the appropriate S function.

2.8 Derive the approximate $(1 - \alpha) \times 100\%$ confidence interval for $S(t)$, the probability of survival beyond time t, given in expression (2.18).

2.9 Derive expression (2.9) by applying the first order Taylor approximation to expression (2.7).

2.10 Derive expression (2.21).

Parametric Methods

Objectives of this chapter:

After studying Chapter 3, the student should:

1. Be familiar with six distributional models.

2. Be able to describe the behavior of their hazard functions.

3. Know that the log-transform of three of these lifetime distributions transforms into a familiar **location and scale family**; and know the relationships between the parameters of the transformed model and those in the original model.

4. Know how to construct a **Q-Q plot** for each of these log(time) distributions.

5. Know the definition of a **likelihood function**.

6. Understand the method of **maximum likelihood estimation** (MLE).

7. Know how to apply the **delta method**.

8. Understand the concept of **likelihood ratio test (LRT)**.

9. Know the general form of the likelihood function for randomly censored data.

10. Understand how to apply the above estimation and testing methods under the exponential model to one sample of data containing censored values. Hence, be familiar with the example of fitting the AML data to an exponential model.

11. Be familiar with the S function `survReg` used to provide a parametric description and analysis of censored data; in particular, how to fit data to the Weibull, log-logistic, and log-normal models.

12. Know how to apply `survReg` to the one-sample and two-sample problems. Be familiar with the additional S functions `anova`, `predict`, and the functions `qq.weibull`, `qq.loglogistic`, `qq.weibreg`, `qq.loglogisreg`, and `qq.lognormreg`, which produce Q-Q plots for one or several samples.

3.1 Frequently used (continuous) models

The exponential distribution

p.d.f. $f(t)$	survivor $S(t)$	hazard $h(t)$
$\lambda \exp(-\lambda t)$	$\exp(-\lambda t)$	$\lambda, \ \lambda > 0$

mean $E(T)$	variance $Var(T)$	pth-quantile t_p
$\frac{1}{\lambda}$	$\frac{1}{\lambda^2}$	$-\lambda^{-1}\log(1-p)$

The outstanding simplicity of this model is its constant hazard rate. We display some p.d.f.'s and survivor functions for three different values of λ in Figure 3.1. The relationship between the cumulative hazard and the survivor

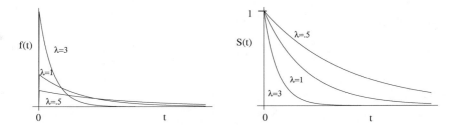

Figure 3.1 *Exponential density and survivor curves.*

function (1.6) is

$$\log\big(H(t)\big) = \log\big(-\log(S(t))\big) = \log(\lambda) + \log(t)$$

or, equivalently expressed with $\log(t)$ on the vertical axis,

$$\boxed{\log(t) = -\log(\lambda) + \log\big(-\log(S(t))\big).} \tag{3.1}$$

Hence, the plot of $\log(t)$ versus $\log\big(-\log(S(t))\big)$ is a straight line with slope 1 and y-intercept $-\log(\lambda)$. At the end of this section we exploit this linear relationship to construct a Q-Q plot for a graphical check of the goodness of fit of the exponential model to the data. Since the hazard function, $h(t) = \lambda$, is constant, plots of both empirical hazards, $\widetilde{h}(t_i)$ and $\widehat{h}(t)$ (page 32), against time provide a quick graphical check. For a good fit, the plot patterns should resemble horizontal lines. Otherwise, look for another survival model. The parametric approach to estimating quantities of interest is presented in Section 3.4. There we first illustrate this with an uncensored sample. Then the same approach is applied to a censored sample. The exponential is a special case of both the Weibull and gamma models, each with their shape parameter equal to 1.

The Weibull distribution

p.d.f. $f(t)$	survivor $S(t)$	hazard $h(t)$
$\lambda\alpha(\lambda t)^{\alpha-1}\times$ $\exp\left(-(\lambda t)^{\alpha}\right)$	$\exp\left(-(\lambda t)^{\alpha}\right)$	$\lambda\alpha(\lambda t)^{\alpha-1}$
mean $E(T)$	variance $Var(T)$	pth-quantile t_p
$\lambda^{-1}\Gamma(1+\frac{1}{\alpha})$	$\lambda^{-2}\Gamma(1+\frac{2}{\alpha})$ $-\lambda^{-2}(\Gamma(1+\frac{1}{\alpha}))^2$	$\lambda^{-1}\left(-\log(1-p)\right)^{\frac{1}{\alpha}}$ $\lambda>0$ and $\alpha>0$

The $\Gamma(k)$ denotes the gamma function and is defined as $\int_0^\infty u^{k-1}e^{-u}du, k>0$. Figure 3.2 displays p.d.f.'s and hazard functions, respectively.

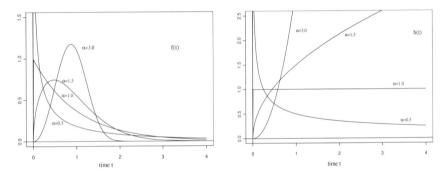

Figure 3.2 *Weibull density and hazard functions with $\lambda=1$.*

Note that the Weibull hazard function is monotone increasing when $\alpha>1$, decreasing when $\alpha<1$, and constant for $\alpha=1$. The parameter α is called the shape parameter as the shape of the p.d.f., and hence the other functions, depends on the value of α. This is clearly seen in Figures 3.2. The λ is a scale parameter in that the effect of different values of λ is just to change the scale on the horizontal (t) axis, not the basic shape of the graph.

This model is very flexible and has been found to provide a good description of many types of time-to-event data. We might expect an increasing Weibull hazard to be useful for modelling survival times of leukemia patients not responding to treatment, where the event of interest is death. As survival time increases for such a patient, and as the prognosis accordingly worsens, the patient's potential for dying of the disease also increases. We might expect some decreasing Weibull hazard to well model the death times of patients recovering from surgery. The potential for dying after surgery usually decreases as the time after surgery increases, at least for a while.

The relationship between the cumulative hazard $H(t)$ and the survivor $S(t)$ (1.6) is seen to be

$$\log\big(H(t)\big) = \log\big(-\log(S(t))\big) = \alpha(\log(\lambda) + \log(t)) \tag{3.2}$$

or equivalently expressed as

$$\boxed{\log(t) = -\log(\lambda) + \sigma \log\big(-\log(S(t))\big)}\,, \tag{3.3}$$

where $\sigma = 1/\alpha$. The plot of $\log(t)$ versus $\log\big(-\log(S(t))\big)$ is a straight line with slope $\sigma = 1/\alpha$ and y-intercept $-\log(\lambda)$. Again, we can exploit this linear relationship to construct a Q-Q plot.

An example of fitting data to the Weibull model using S, along with its Q-Q plot, is presented in Section 3.4. This distribution is intrinsically related to the extreme value distribution which is the next distribution to be discussed. The natural log transform of a Weibull random variable produces an extreme value random variable. This relationship is exploited quite frequently, particularly in the statistical computing packages and in diagnostic plots.

The extreme (minimum) value distribution

The interest in this distribution is not for its direct use as a lifetime distribution, but rather because of its relationship to the Weibull distribution. Let μ, where $-\infty < \mu < \infty$, and $\sigma > 0$ denote location and scale parameters, respectively. The standard extreme value distribution has $\mu = 0$ and $\sigma = 1$.

p.d.f. $f(y)$	survivor $S(y)$	
$\sigma^{-1}\exp\left(\frac{y-\mu}{\sigma} - \exp\left(\frac{y-\mu}{\sigma}\right)\right)$	$\exp\left(-\exp\left(\frac{y-\mu}{\sigma}\right)\right)$	
mean $E(Y)$	variance $Var(Y)$	pth-quantile y_p
$\mu - \gamma\sigma$	$\frac{\pi^2}{6}\sigma^2$	$y_p = \mu$ $+\sigma\log\big(-\log(1-p)\big)$

Here γ denotes Euler's constant, $\gamma = 0.5772...$, the location parameter μ is the 0.632th quantile, and y can also be negative so that $-\infty < y < \infty$. Further, the following relationship can be easily shown:

Fact: If T is a Weibull random variable with parameters α and λ, then $Y = \log(T)$ follows an extreme value distribution with $\mu = -\log(\lambda)$ and $\sigma = \alpha^{-1}$. The r.v. Y can be represented as $Y = \mu + \sigma Z$, where Z is a standard extreme value r.v., as the extreme value distribution is a location and scale family of distributions.

As values of μ and σ different from 0 and 1 do not effect the shape of the p.d.f., but only location and scale, displaying only plots of the standard extreme value p.d.f. and survivor function in Figure 3.3 suffices.

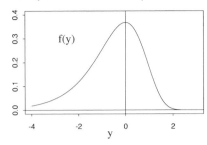 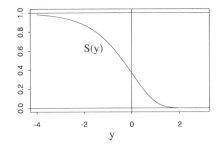

Figure 3.3 *Standard extreme value density and survivor functions.*

The log-normal distribution

This distribution is most easily characterized by saying the lifetime T is log-normally distributed if $Y = \log(T)$ is normally distributed with mean and variance specified by μ and σ^2, respectively. Hence, Y is of the form $Y = \mu + \sigma Z$ where Z is a standard normal r.v. We have the following table for T with $\alpha > 0$ and $\lambda > 0$ and where $\Phi(\cdot)$ denotes the standard normal d.f.:

p.d.f. $f(t)$	survivor $S(t)$	hazard $h(t)$
$(2\pi)^{-\frac{1}{2}}\alpha t^{-1}\exp\left(\frac{-\alpha^2(\log(\lambda t))^2}{2}\right)$	$1 - \Phi\left(\alpha\log(\lambda t)\right)$	$\frac{f(t)}{S(t)}$
mean $E(T)$	**variance $Var(T)$**	**Note:**
$\exp(\mu + \frac{\sigma^2}{2})$	$(\exp(\sigma^2) - 1)\times$ $\exp(2\mu + \sigma^2)$	$\mu = -\log(\lambda)$ and $\sigma = \alpha^{-1}$

The hazard function has value 0 at $t = 0$, increases to a maximum, and then decreases, approaching zero as t becomes large. Since the hazard decreases for large values of t, it seems implausible as a lifetime model in most situations. But, it can still be suitable for representing lifetimes, particularly when large values of t are not of interest. We might also expect this hazard to describe tuberculosis patients well. Their potential for dying increases early in the disease and decreases later. Lastly, the log-logistic distribution, to be presented next, is known to be a good approximation to the log-normal and is often a preferred survival time model. Some p.d.f's and hazard functions are displayed in Figure 3.4.

The log-logistic distribution

The lifetime T is log-logistically distributed if $Y = \log(T)$ is logistically distributed with location parameter μ and scale parameter σ. Hence, Y is also

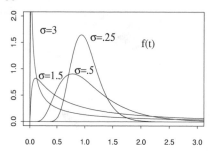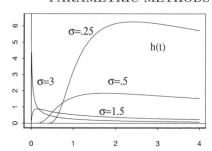

Figure 3.4 *Log-normal densities and hazards with $\mu = 0$ and $\sigma = .25, .5, 1.5,$ and 3.*

of the form $Y = \mu + \sigma Z$ where Z is a standard logistic r.v. with density

$$\frac{\exp(z)}{\left(1 + \exp(z)\right)^{2}}, \quad -\infty < z < \infty.$$

This is a symmetric density with mean 0 and variance $\pi^2/3$, and with slightly heavier tails than the standard normal, the excess in kurtosis being 1.2. We have the following table for T with $\alpha > 0$ and $\lambda > 0$:

p.d.f. $f(t)$	survivor $S(t)$	hazard $h(t)$
$\lambda\alpha(\lambda t)^{\alpha-1}\left(1 + (\lambda t)^{\alpha}\right)^{-2}$	$\frac{1}{1+(\lambda t)^{\alpha}}$	$\frac{\lambda\alpha(\lambda t)^{\alpha-1}}{1+(\lambda t)^{\alpha}}$
Note:		pth-quantile t_p
$\mu = -\log(\lambda)$ and $\sigma = \alpha^{-1}$		$\lambda^{-1}\left(\frac{p}{1-p}\right)^{\frac{1}{\alpha}}$

This model has become popular, for like the Weibull, it has simple algebraic expressions for the survivor and hazard functions. Hence, handling censored data is easier than with the log-normal while providing a good approximation to it except in the extreme tails. The hazard function is identical to the Weibull hazard aside from the denominator factor $1 + (\lambda t)^{\alpha}$. For $\alpha < 1$ ($\sigma > 1$) it is monotone decreasing from ∞ and is monotone decreasing from λ if $\alpha = 1$ ($\sigma = 1$). If $\alpha > 1$ ($\sigma < 1$), the hazard resembles the log-normal hazard as it increases from zero to a maximum at $t = (\alpha - 1)^{1/\alpha}/\lambda$ and decreases toward zero thereafter. In Section 3.4 an example of fitting data to this distribution using S along with its Q-Q plot is presented. Some p.d.f.'s and hazards are displayed in Figure 3.5.

We exploit the simple expression for the survivor function to obtain a relationship which is used for checking the goodness of fit of the log-logistic model to the data. The odds of survival beyond time t are

$$\frac{S(t)}{1 - S(t)} = (\lambda t)^{-\alpha}. \tag{3.4}$$

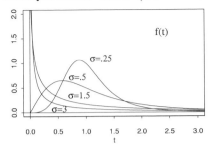 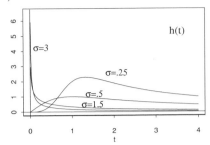

Figure 3.5 *Log-logistic densities and hazards with* $\mu = 0$ *and* $\sigma = .25, .5, 1.5,$ *and* 3.

It easily follows that $\log(t)$ is a linear function of the log-odds of survival beyond t. The precise linear relationship is

$$\log(t) = \mu + \sigma\left(-\log\left(\tfrac{S(t)}{1-S(t)}\right)\right), \tag{3.5}$$

where $\mu = -\log(\lambda)$ and $\sigma = 1/\alpha$. The plot of the $\log(t)$ against $-\log\{S(t)/(1-S(t))\}$ is a straight line with slope σ and y-intercept μ. At the end of this section, the Q-Q plot is constructed using this linear relationship.

The gamma distribution

Like the Weibull, this distribution has a scale parameter $\lambda > 0$ and shape parameter $k > 0$ and contains the exponential distribution as a special case; i.e., when shape $k = 1$. As a result, this model is also more flexible than the exponential. We have the following table for this distribution:

p.d.f. $f(t)$	survivor $S(t)$	hazard $h(t)$
$\frac{\lambda^k t^{k-1}}{\Gamma(k)}\exp(-\lambda t)$	no simple form	no simple form
mean $E(T)$	variance $Var(T)$	
$\frac{k}{\lambda}$	$\frac{k}{\lambda^2}$	

The hazard function is monotone increasing from 0 when $k > 1$, monotone decreasing from ∞ if $k < 1$, and in either case approaches λ as t increases.

The model for $Y = \log(T)$ can be written $Y = \mu + Z$, where Z has density

$$\frac{\exp\left(kz - \exp(z)\right)}{\Gamma(k)}. \tag{3.6}$$

The r.v. Y is called a log-gamma r.v. with parameters k and $\mu = -\log(\lambda)$. The quantity Z has a negatively skewed distribution with skewness decreasing with k increasing. When $k = 1$, this is the exponential model and, hence, Z has the standard extreme value distribution. With the exception of $k = 1$, the

log-gamma is not a member of the location and scale family of distributions.
It is, however, a member of the location family. Figure 3.6 shows some gamma
p.d.f.'s and hazards. We display some log-gamma p.d.f.'s in Figure 3.7. See
Klein & Moeschberger (1997, page 44) and Kalbfleisch & Prentice (1980, page
27) for a discussion of the generalized gamma and corresponding generalized
log-gamma distributions.

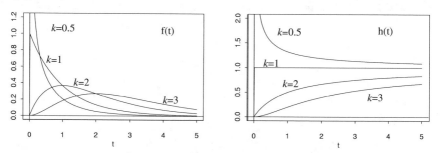

Figure 3.6 *Gamma densities and hazards with* $\lambda = 1$ *and* $k = 0.5, 1, 2,$ *and 3.*

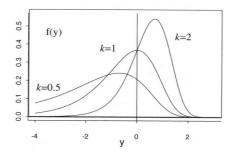

Figure 3.7 *Log-gamma density with* $k = 0.5, 1, 2,$ *and* $\lambda = 1.$

A note on discrete lifetimes

When T is discrete r.v., different techniques are required. Discrete random
variables in survival analyses arise due to rounding off measurements, grouping
of failure times into intervals, or when lifetimes refer to an integral number
of units. When effectively continuous data have been grouped into intervals,
the multinomial distribution is used. Methods for multinomial models are
discussed in Lawless (1982, Chapter 2). Agresti (1990, Chapter 6.6) presents
methods to analyze mortality rates and survival times using log-linear models
when explanatory variables are categorical.

Summary

Except for the gamma distribution, all distributions of lifetime T we work
with have the property that the distribution of the log-transform $\log(T)$ is a

member of the location and scale family of distributions. The common features are:

- The time T distributions have two parameters –

$$\text{scale} = \lambda \quad \text{and} \quad \text{shape} = \alpha \, .$$

- In log-time, $Y = \log(T)$, the distributions have two parameters –

$$\text{location} = \mu = -\log(\lambda) \quad \text{and} \quad \text{scale} = \sigma = \frac{1}{\alpha} \, .$$

- Each can be expressed in the form

$$Y = \log(T) = \mu + \sigma Z \, , \tag{3.7}$$

where Z is the standard member; that is,

$$\mu = 0 \; (\lambda = 1) \quad \text{and} \quad \sigma = 1 \; (\alpha = 1) \, .$$

- They are log-linear models.

The three distributions considered in our examples are summarized as follows:

T	$\Longleftrightarrow$	$Y = \log(T)$
Weibull	$\Longleftrightarrow$	extreme minimum value
log-normal	$\Longleftrightarrow$	normal
log-logistic	$\Longleftrightarrow$	logistic

If the true distribution of $Y = \log(T)$ is one of the above, then the pth-quantile y_p is a linear function of z_p, the pth-quantile of the standard member of the specified distribution. The straight line has slope σ and y-intercept μ. Let t_p denote an arbitrary pth-quantile. In light of the foregoing discussion, the linear relationships for $y_p = \log(t_p)$ reported in expressions (3.3), (3.5), (3.7) take on new meaning. This is summarized in Table 3.1.

Construction of the quantile-quantile (Q-Q) plot

Let $\widehat{S}(t)$ denote the K-M estimator of survival probability beyond time t. Let $t_i, i = 1, \ldots, r \leq n$, denote the ordered uncensored observed failure times. For each uncensored sample quantile $y_i = \log(t_i)$, the estimated failure probability is $\hat{p}_i = 1 - \widehat{S}(t_i)$. The parametric standard quantile z_i is obtained by using the $\hat{p}_i$ to evaluate the expression for the standard quantile given in Table 3.1. Thus, $F_{0,1}(z_i) = P(Z \leq z_i) = \hat{p}_i$, where $F_{0,1}$ is the d.f. of the standard parametric model ($\mu = 0, \sigma = 1$) under consideration. As the K-M estimator is distribution free and consistently estimates the "true" survival function, for large sample sizes n, the z_i should reflect the "true" standard quantiles, if F is indeed the "true" lifetime d.f.. Hence, if the proposed model fits the data

Table 3.1: *Relationships to exploit to construct a graphical check for model adequacy*

t_p quantile	$y_p = \log(t_p)$ quantile	form of standard quantile z_p
Weibull	extreme value	$\log(-\log(S(t_p))) = \log(H(t_p))$ $= \log(-\log(1-p))$
log-normal	normal	$\Phi^{-1}(p)$, where Φ denotes the standard normal d.f.
log-logistic	logistic	$-\log\left(\dfrac{S(t_p)}{1 - S(t_p)}\right) = -\log(\text{odds})$ $= -\log\left(\frac{1-p}{p}\right)$

adequately, the points (z_i, y_i) should lie close to a straight line with slope σ and y-intercept μ. **The plot of the points (z_i, y_i) is called a quantile-quantile (Q-Q) plot**. An appropriate line to compare the plot pattern to is one with the maximum likelihood estimates $\hat{\sigma}$ and $\hat{\mu}$ to be discussed in the next section. Plot patterns grossly different from this straight line indicate the proposed model is inadequate. The more closely the plot pattern follows this line, the more evidence there is in support of the proposed model. If the model is deemed appropriate, the slope and y-intercept of an "empirical" line, obtained from least squares or some robust procedure, provide rough estimates of $\sigma = 1/\alpha$ and $\mu = -\log(\lambda)$, respectively. The Q-Q plot is a major diagnostic tool for checking model adequacy.

A cautionary note: Fitting the uncensored points (z_i, y_i) to a least squares line alone can be very misleading in deeming model adequacy. Our first example of this is discussed in Section 3.4, where we first construct Q-Q plots to check and compare the adequacy of fitting the AML data to the exponential, Weibull, and log-logistic distributions.

3.2 Maximum likelihood estimation (MLE)

Our assumptions here are that the $T_1, \ldots, T_n$ are iid from a continuous distribution with p.d.f. $f(t|\theta)$, where θ belongs to some parameter space Ω. Here, θ could be either a real-valued or vector-valued parameter. The **likelihood function** is the joint p.d.f. of the sample when regarded as a function of θ for a given value $(t_1, \ldots, t_n)$. To emphasize this we denote it by $L(\theta)$. For a random sample, this is the product of the p.d.f.'s. That is, the likelihood function is given by

$$L(\theta) = \prod_{i=1}^{n} f(t_i|\theta).$$

The **maximum likelihood estimator** (MLE), denoted by $\widehat{\theta}$, is the value of θ in Ω that maximizes $L(\theta)$ or, equivalently, maximizes the log-likelihood

$$\log L(\theta) = \sum_{i=1}^{n} \log f(t_i|\theta).$$

MLE's possess the *invariance property*; that is, the MLE of a function of θ, say $\tau(\theta)$, is $\tau(\widehat{\theta})$. For a gentle introduction to these foregoing notions, see DeGroot (1986). Under the random censoring model, we see from expression (1.13) that if we assume that the censoring time has no connection to the survival time, then the log-likelihood for the maximization process can be taken to be

$$\log L(\theta) = \log \prod_{i=1}^{n} f^{\delta_i}(y_i|\theta) S_f^{1-\delta_i}(y_i|\theta) = \sum_u \log f(y_i|\theta) + \sum_c \log S_f(y_i|\theta),$$

(3.8)

where u and c mean sums over the uncensored and censored observations, respectively. Let $I(\theta)$ denote the **Fisher information matrix**. Then its elements are

$$I(\theta) = \left(\left(-E(\frac{\partial^2}{\partial\theta_j\partial\theta_k} \log L(\theta)) \right) \right),$$

where E denotes expectation. As we are working with random samples (iid) we point out that $I(\theta)$ can be expressed as

$$I(\theta) = nI_1(\theta),$$

where $I_1(\theta) = \left(\left(-E(\frac{\partial^2}{\partial\theta_j\partial\theta_k} \log f(y_1|\theta)) \right) \right)$ is the Fisher information matrix of any one of the observations.

The MLE $\widehat{\theta}$ has the following **large sample distribution**:

$$\widehat{\theta} \overset{a}{\sim} \mathbf{MVN}(\theta, I^{-1}(\theta)),$$

(3.9)

where MVN denotes multivariate normal and $\overset{a}{\sim}$ is read "is asymptotically distributed." The asymptotic covariance matrix $I^{-1}(\theta)$ is a $d \times d$ matrix, where d is the dimension of θ. The ith diagonal element of $I^{-1}(\theta)$ is the asymptotic variance of the ith component of $\widehat{\theta}$. The off-diagonal elements are the asymptotic covariances of the corresponding components of $\widehat{\theta}$. If θ is a scalar (real valued), then the asymptotic variance, denoted var_a, of $\widehat{\theta}$ is

$$\text{var}_a(\hat{\theta}) = \frac{1}{I(\theta)},$$

where $I(\theta) = -E\left(\partial^2 \log L(\theta)/\partial\theta^2\right)$. For censored data, this expectation is a function of the censoring distribution G as well as the survival time distribution F. Hence, it is necessary to approximate $I(\theta)$ by the **observed information matrix** $i(\theta)$ evaluated at the MLE $\widehat{\theta}$, where

$$i(\theta) = \left(\left(-\frac{\partial^2}{\partial\theta_j\partial\theta_k} \log L(\theta) \right) \right).$$

(3.10)

For the univariate case,

$$i(\theta) = -\frac{\partial^2 \log L(\theta)}{\partial \theta^2}. \tag{3.11}$$

Hence, $\mathrm{var}_a(\hat{\theta})$ is approximated by $\left(i(\hat{\theta})\right)^{-1}$.

The **delta method** is useful for obtaining limiting distributions of smooth functions of an MLE. When variance of an estimator includes the parameter of interest, the delta method can be used to remove the parameter in the variance. This is called the variance-stabilization. We describe it for the univariate case.

Delta method:

Suppose a random variable Z has a mean μ and variance σ^2 and suppose we want to approximate the distribution of some function $g(Z)$. Take a first order Taylor expansion of $g(Z)$ about μ and ignore the higher order terms to get

$$g(Z) \approx g(\mu) + (Z - \mu)g'(\mu).$$

Then the mean$(g(Z)) \approx g(\mu)$ and the var$(g(Z)) \approx (g'(\mu))^2 \sigma^2$. Furthermore, if

$$Z \overset{a}{\sim} \mathrm{normal}(\mu, \sigma^2),$$

then

$$g(Z) \overset{a}{\sim} \mathrm{normal}(g(\mu), (g'(\mu))^2 \sigma^2). \tag{3.12}$$

Example: Let $X_1, \ldots, X_n$ be iid from a Poisson distribution with mean λ. Then the MLE of λ is $\hat{\lambda} = \overline{X}$. We know that the mean and variance of $Z = \overline{X}$ are λ and λ/n, respectively. Take $g(Z) = \overline{X}^{\frac{1}{2}}$. Then $g(\lambda) = \lambda^{\frac{1}{2}}$ and

$$\overline{X}^{\frac{1}{2}} \overset{a}{\sim} \mathrm{normal} \text{ with mean} \approx \lambda^{\frac{1}{2}} \text{ and variance} \approx \frac{1}{4n}.$$

There are multivariate versions of the delta method. One is stated in Section 3.6.

3.3 Confidence intervals and tests

For some estimators we can compute their small sample exact distributions. However, for most, in particular when censoring is involved, we must rely on the large sample properties of the MLE's. For confidence intervals or for testing $H_0 : \theta = \theta_0$, where θ is a scalar or a scalar component of a vector, we can construct the asymptotic z-intervals with the standard errors (s.e.) taken from the diagonal of the asymptotic covariance matrix which is the inverse of the information matrix $I(\theta)$ evaluated at the MLE $\hat{\theta}$ if necessary. The s.e.'s are, of course, the square roots of these diagonal values. In summary:

An approximate $(1 - \alpha) \times 100\%$ confidence interval for the parameter θ is given by

$$\hat{\theta} \pm z_{\frac{\alpha}{2}} \text{s.e.}(\hat{\theta}), \tag{3.13}$$

where $z_{\frac{\alpha}{2}}$ is the upper $\frac{\alpha}{2}$ quantile of the standard normal distribution and, by (3.11), s.e. is the square root of $\text{var}_a(\hat{\theta}) \approx -\left(\partial^2 \log L(\theta)/\partial \theta^2\right)^{-1} = \left(i(\widehat{\theta})\right)^{-1}$.

However, if we are performing joint estimation or testing a vector-valued θ, we have three well known procedures: Assume θ_0 has d-components, $d \geq 1$. Unless otherwise declared, $\widehat{\theta}$ denotes the MLE.

- The **Wald** statistic:

$$(\widehat{\theta} - \theta_0)' I(\theta_0)(\widehat{\theta} - \theta_0) \overset{a}{\sim} \chi^2_{(d)} \text{ under } H_0.$$

- The **Rao** statistic:

$$\frac{\partial}{\partial \theta} \log L(\theta_0)' I^{-1}(\theta_0) \frac{\partial}{\partial \theta} \log L(\theta_0) \overset{a}{\sim} \chi^2_{(d)} \text{ under } H_0.$$

Note that Rao's method does not use the MLE. Hence, no iterative calculation is necessary.

- The Neyman-Pearson/Wilks **likelihood ratio test** (**LRT**): Let the vector $\underline{t}$ represent the n observed values; that is, $\underline{t}' = (t_1, \ldots, t_n)$. The LRT statistic is given by

$$r^*(\underline{t}) = -2 \log \left(\frac{L(\theta_0)}{L(\widehat{\theta})} \right) \overset{a}{\sim} \chi^2_{(d)} \text{ under } H_0. \tag{3.14}$$

To test $H_0 : \theta = \theta_0$ against $H_A : \theta \neq \theta_0$, we reject for small values of $L(\theta_0)/L(\widehat{\theta})$ (as this ratio is less than or equal to 1). Equivalently, we reject for large values of $r^*(\underline{t})$.

For **joint confidence regions** we simply take the region of values that satisfy the elliptical region formed with either the Wald or Rao statistic with $I(\theta)$ or $i(\theta)$ evaluated at the MLE $\widehat{\theta}$. For example, an approximate $(1 - \alpha) \times 100\%$ joint confidence region for θ is given by

$$\{\theta; \text{Wald} \leq \chi^2_\alpha\},$$

where χ^2_α is the chi-square upper αth-quantile with d degrees of freedom. The following picture explains:

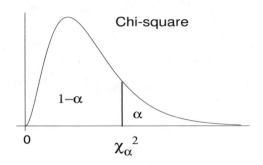

3.4 One-sample problem

3.4.1 Fitting data to the exponential model

Case 1: No censoring.

All "failures" are observed. The $T_1, \ldots, T_n$ are iid.

- **Likelihood:**

$$L(\lambda) = \prod_{i=1}^{n} \lambda \exp(-\lambda t_i) = \lambda^n \exp\left(-\lambda \sum_{i=1}^{n} t_i\right)$$

- **Log-likelihood:**

$$\log L(\lambda) = n \log \lambda - \lambda \sum_{i=1}^{n} t_i$$

$$\frac{\partial \log L(\lambda)}{\partial \lambda} = \frac{n}{\lambda} - \sum_{i=1}^{n} t_i$$

- **MLE:**
  Set

$$\frac{\partial \log L(\lambda)}{\partial \lambda} = 0$$

  and solve for λ. Therefore,

$$\widehat{\lambda} = \frac{n}{\sum_{i=1}^{n} t_i} = \frac{1}{\overline{T}}.$$

  The MLE of the mean $\theta = 1/\lambda$ is $\widehat{\theta} = \overline{T}$.

- **Exact distribution theory:**
  Since the T_i are iid exponential(λ), the sum $\sum_{i=1}^{n} T_i$ has a gamma distribution with parameters $k = n$ and λ. From basic theory, we know

$$2\lambda \sum_{i=1}^{n} T_i = 2n\frac{\lambda}{\widehat{\lambda}} \sim \chi^2_{(2n)}. \tag{3.15}$$

WHY! This can be used as a pivotal quantity to construct both test and confidence interval.

- **A $(1 - \alpha) \times 100\%$ confidence interval (C.I.) for λ** follows from the next picture:

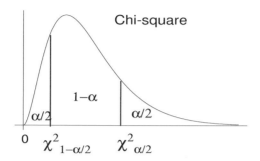

With probability $1-\alpha$,

$$\chi^2_{1-\alpha/2} \leq \frac{2n\lambda}{\widehat{\lambda}} \leq \chi^2_{\alpha/2} \ .$$

It then follows from simple algebra that: A $(1 - \alpha) \times 100\%$ C. I. for λ is given by

$$\frac{\widehat{\lambda}}{2n}\chi^2_{1-\alpha/2} \leq \lambda \leq \frac{\widehat{\lambda}}{2n}\chi^2_{\alpha/2} \ .$$

Let $\theta = 1/\lambda$. Then θ is the mean of the distribution. It follows:

A $(1 - \alpha) \times 100\%$ C.I. for the mean survival time θ is given by

$$\frac{2n\overline{T}}{\chi^2_{\alpha/2}} \leq \theta \leq \frac{2n\overline{T}}{\chi^2_{1-\alpha/2}} \ .$$

Or, simply invert the endpoints of the C.I. for λ and reverse the inequalities.

- **The pth-quantile is the t_p** such that $F(t_p|\lambda) = p$. Thus t_p is such that $1 - \exp(-\lambda t_p) = p$. Therefore, $t_p = -\log(1 - p)/\lambda$. By the invariance property of MLE's, the MLE of t_p is

$$\widehat{t}_p = -\widehat{\lambda}^{-1} \log(1 - p) = -\overline{T} \log(1 - p) \ .$$

Then the MLE of the median is

$$\widehat{\mathrm{med}} = -\overline{T} \log(0.5) \ .$$

We pretend there are no censored points and use the AML data to illustrate the various calculations. We work with the maintained group's data values.

$$n = 11, \qquad \sum_{i=1}^{n} t_i = 423 \qquad \text{degrees of freedom} = 2 \cdot 11 = 22$$

MLE's : $\widehat{\theta} = \overline{t} = 38.4545,$ $\widehat{\lambda} = 0.026.$

For a 95% confidence interval, $\chi^2_{0.025,22} = 36.78,$ $\chi^2_{0.975,22} = 10.98.$

- **A 95% C.I. for λ is given by**

$$\frac{0.026}{2 \cdot 11} \cdot 10.98 \leq \lambda \leq \frac{0.026}{2 \cdot 11} \cdot 36.78 \quad \Longleftrightarrow \quad 0.01298 \leq \lambda \leq 0.04347 \,.$$

- **A 95% C.I. for θ**, the mean survival (in weeks), is given by

$$\frac{2 \cdot 423}{36.78} \leq \theta \leq \frac{2 \cdot 423}{10.98} \quad \Longleftrightarrow \quad 23 \leq \theta \leq 77.05 \,.$$

Or, simply invert the endpoints of the previous interval and reverse the inequalities.

- **The MLE of median:**

$$\widehat{\text{med}} = -\overline{T}\log(0.5) = -38.4545 \cdot \log(0.5) = 26.6546 \text{ weeks} \quad < \quad \overline{T} \,.$$

- **To test**

$$H_0 : \text{ mean } \theta = 30 \text{ weeks} \quad \Longleftrightarrow \quad (\lambda = 1/30 = 0.03\overline{3})$$

against

$$H_A : \theta \neq 30 \quad \Longleftrightarrow \quad (\lambda \neq 0.03\overline{3})$$

at the 5% level of significance, we can use the exact confidence interval for θ obtained above. We reject H_0 if the 95% confidence interval does not contain 30. Therefore, we do not reject H_0. The mean survival is not significantly different from 30 weeks. For a one-sided test, the significance-level would be 2.5%. We can base a test on the test-statistic

$$T^* = 2\lambda_0 \sum_{i=1}^{n} T_i \sim \chi^2_{(2n)} \quad \text{under} \quad H_0 : \lambda = \lambda_0 \,.$$

To test against $H_A : \lambda \neq \lambda_0$, construct a two-tailed size α critical region. Here

$$t^* = 2 \cdot 0.03\overline{3} \cdot 423 = 28.2 \,.$$

At $\alpha = .05$, $df = 22$, $\chi^2_{.975} = 10.98$, and $\chi^2_{.025} = 36.78$, we fail to reject H_0. This is a flexible test as you can test one-sided alternatives. For example, to test $H_A : \lambda < \lambda_0$ ($\theta > \theta_0$), compute the p-value. Thus,

$$p\text{-value} = P(T^* \geq 28.2) = 0.17 \,.$$

Again, we fail to reject H_0. The p-value for the two-sided alternative is then 0.34.

- **The likelihood ratio test (LRT)** (3.14)

The LRT can be shown to be equivalent to the two-sided test based on the test statistic T^* just above. Therefore, here we will use the asymptotic distribution and then compare. The test statistic

$$r^*(\underline{t}) = -2\log\left(\frac{L(\lambda_0)}{L(\widehat{\lambda})}\right) \overset{a}{\sim} \chi^2_{(1)}.$$

We reject $H_0 : \theta = 30$ weeks when $r^*(\underline{t})$ is large.

$$
\begin{aligned}
r^*(\underline{t}) &= -2\log L(\lambda_0) + 2\log L(\widehat{\lambda}) \\
&= -2n\log(\lambda_0) + 2\lambda_0 n\overline{t} + 2n\log(1/\overline{t}) - 2n \\
&= -2\cdot 11\cdot\log(1/30) + \frac{2}{30}\cdot 423 + 2\cdot 11\cdot\log(11/423) - 2\cdot 11 \\
&= 0.7378
\end{aligned}
$$

The p-value $= P(r^*(\underline{t}) \ge 0.7378) \approx 0.39$. Therefore, we fail to reject H_0. This p-value is very close to the exact p-value 0.34 computed above.

Case 2: Random censoring.

Let u, c, and n_u denote uncensored, censored, and number of uncensored observations, respectively. The n observed values are now represented by the vectors $\underline{y}$ and $\underline{\delta}$, where $\underline{y}' = (y_1, \dots, y_n)$ and $\underline{\delta}' = (\delta_1, \dots, \delta_n)$. Then

- **Likelihood:** See expressions (1.13), (3.8).

$$
\begin{aligned}
L(\lambda) &= \prod_u f(y_i|\lambda) \cdot \prod_c S_f(y_i|\lambda) \\
&= \prod_u \lambda\exp(-\lambda y_i)\prod_c \exp(-\lambda y_i) \\
&= \lambda^{n_u}\exp\left(-\lambda\sum_u y_i\right)\exp\left(-\lambda\sum_c y_i\right) \\
&= \lambda^{n_u}\exp\left(-\lambda\sum_{i=1}^n y_i\right)
\end{aligned}
$$

- **Log-likelihood:**

$$\log L(\lambda) = n_u\log(\lambda) - \lambda\sum_{i=1}^n y_i$$

$$\frac{\partial\log L(\lambda)}{\partial\lambda} = \frac{n_u}{\lambda} - \sum_{i=1}^n y_i$$

$$\frac{\partial^2\log L(\lambda)}{\partial\lambda^2} = -\frac{n_u}{\lambda^2} = -i(\lambda),\text{ the negative of the observed information.}$$

- **MLE:**

$$\widehat{\lambda} = \frac{n_u}{\sum_{i=1}^n y_i} \qquad\text{and}\qquad \mathrm{var}_a(\widehat{\lambda}) = \left(-E\left(-\frac{n_u}{\lambda^2}\right)\right)^{-1} = \frac{\lambda^2}{E(n_u)},$$

where $E(n_u) = n \cdot P(T \leq C)$. From expression (3.9),

$$\frac{\widehat{\lambda} - \lambda}{\sqrt{\lambda^2/E(n_u)}} \overset{a}{\sim} N(0, 1).$$

We replace $E(n_u)$ by n_u since we don't usually know the censoring distribution $G(\cdot)$. Notice the dependence of the asymptotic variance on the unknown parameter λ. We substitute in $\widehat{\lambda}$ and obtain

$$\mathrm{var}_a(\widehat{\lambda}) \approx \frac{\widehat{\lambda}^2}{n_u} = \frac{1}{i(\widehat{\lambda})},$$

where $i(\lambda)$ is just above. The MLE for the mean $\theta = 1/\lambda$ is simply $\widehat{\theta} = 1/\widehat{\lambda} = \sum_{i=1}^n y_i/n_u$.

On the AML data, $n_u = 7$,

$$\widehat{\lambda} = \frac{7}{423} = 0.0165, \quad \text{and} \quad \mathrm{var}_a(\widehat{\lambda}) \approx \frac{\widehat{\lambda}^2}{7} = \frac{0.0165^2}{7}.$$

- **A 95% C.I. for λ** (3.13) is given by

$$\widehat{\lambda} \pm z_{0.025} \cdot \mathrm{se}(\widehat{\lambda}) =: \ 0.0165 \pm 1.96 \cdot \frac{0.0165}{\sqrt{7}} =: \ [0.004277 \ , \ 0.0287].$$

- **A 95% C.I. for θ**, the mean survival, can be obtained by inverting the previous interval for λ. This interval is: $[34.8, 233.808]$ weeks. Both intervals are very skewed. However, as $\widehat{\theta} = 1/\widehat{\lambda} = 60.42856$ weeks, we have $\theta = g(\lambda) = 1/\lambda$ and we can use the delta method to obtain the asymptotic variance of $\widehat{\theta}$. As $g'(\lambda) = -\lambda^{-2}$, the asymptotic variance is

$$\mathrm{var}_a(\widehat{\theta}) = \frac{1}{\lambda^2 E(n_u)} \approx \frac{1}{\widehat{\lambda}^2 \cdot n_u} = \frac{\widehat{\theta}^2}{n_u}. \tag{3.16}$$

Hence, a second 95% C.I. for θ, the mean survival, is given by

$$\widehat{\theta} \pm z_{0.025} \mathrm{se}(\widehat{\theta}) =: \ 60.42856 \pm 1.96 \cdot \frac{1}{0.0165 \cdot \sqrt{7}} =: \ [15.66246, 105.1947] \ \text{weeks}.$$

Notice this is still skewed, but much less so; and it is much narrower. Here we use the asymptotic variance of $\widehat{\theta}$ directly, and hence, eliminate one source of variation. However, the asymptotic variance still depends on λ.

- **The MLE of the pth-quantile:**

$$\widehat{t_p} = -\frac{1}{\widehat{\lambda}} \log(1 - p) = -\frac{\sum_{i=1}^n y_i}{n_u} \log(1 - p).$$

Thus, the MLE of the median is

$$\widehat{\mathrm{med}} = -\frac{423}{7} \log(0.5) = 41.88 \ \text{weeks}.$$

Notice how much smaller the median is compared to the estimate $\widehat{\theta} =$

60.43. The median reflects a more typical survival time. The mean is greatly influenced by the one large value 161+. Note that

$$\text{var}_a(\widehat{t}_p) = \Big(\log(1-p)\Big)^2 \cdot \text{var}_a\Big(\widehat{\lambda}^{-1}\Big) \approx \Big(\log(1-p)\Big)^2 \cdot \frac{1}{\widehat{\lambda}^2 \cdot n_u}.$$

The $\text{var}_a\Big(\widehat{\lambda}^{-1}\Big)$ is given in expression (3.16). Thus, **a 95% C.I. for the median** is given by

$$\widehat{t}_{0.5} \pm 1.96 \cdot \frac{-\log(0.5)}{\widehat{\lambda} \cdot \sqrt{n_u}} =: \quad 41.88 \pm 1.96 \cdot \frac{-\log(0.5)}{0.0165 \cdot \sqrt{7}} =: \quad [10.76, 73] \text{ weeks.}$$

- With the delta method (3.12) we can construct intervals that are less skewed and possibly narrower by finding transformations which eliminate the dependence of the asymptotic variance on the unknown parameter of interest. For example, the natural log-transform of $\widehat{\lambda}$ accomplishes this. This is because for $g(\lambda) = \log(\lambda)$, $g'(\lambda) = 1/\lambda$ and, thus, $\text{var}_a(\log(\widehat{\lambda})) = \lambda^{-2}\{\lambda^2/E(n_u)\} = 1/E(n_u)$. Again we replace $E(n_u)$ by n_u. Therefore, we have

$$\log(\widehat{\lambda}) \overset{a}{\sim} N\left(\log(\lambda), \frac{1}{n_u}\right). \tag{3.17}$$

A 95% C.I. for $\log(\lambda)$ is given by

$$\log(\widehat{\lambda}) \pm 1.96 \cdot \frac{1}{\sqrt{n_u}}$$

$$\log\left(\frac{7}{423}\right) \pm 1.96 \cdot \frac{1}{\sqrt{7}}$$

$$[-4.84, -3.36].$$

Transform back by taking exp(endpoints). This second 95% C.I. for λ is

$$[.0079, .0347],$$

which is slightly wider than the previous interval for λ. Invert and reverse endpoints to obtain a third 95% C.I. for the mean θ. This yields $[28.81, 126.76]$ weeks, which is also slightly wider than the second interval for θ.

Analogously, since $\text{var}_a(\widehat{\theta}) \approx \widehat{\theta}^2/n_u$ (3.16), the delta method provides large sample distributions for $\log(\widehat{\theta})$ and $\log(\widehat{t}_p)$ with the same variance, which is free of the parameter θ. They are

$$\log(\widehat{\theta}) \overset{a}{\sim} N\left(\log(\theta), \frac{1}{n_u}\right) \tag{3.18}$$

$$\log(\widehat{t}_p) \overset{a}{\sim} N\left(\log(t_p), \frac{1}{n_u}\right). \tag{3.19}$$

Analogously, we first construct C.I.'s for the log(parameter), then take exp(endpoints) to obtain C.I.'s for the parameter. Most statisticians prefer this approach. Using the AML data, we summarize 95% C.I.'s in Table 3.2.

Table 3.2: *Preferred 95% confidence intervals for mean and median (or any quantile) of an exponential survival model based on the log-transform*

parameter	point estimate	log(parameter)	parameter
mean	60.43 weeks	[3.361, 4.84]	[28.8₹, 126.76] weeks
median	41.88 weeks	[2.994, 4.4756]	[19.965, 87.85] weeks

e e

- **The MLE of the survivor function $S(t) = \exp(-\lambda t)$:**

$$\widehat{S}(t) = \exp(-\widehat{\lambda}t) = \exp(-0.0165\,t).$$

For any fixed t, $\widehat{S}(t)$ is a function of $\widehat{\lambda}$. We can get its approximate distribution by using the delta method. Alternatively, we can take a log-log transformation that usually improves the convergence to normality. This is because the var_a is free of the unknown parameter λ. This follows from (3.17) and the relationship

$$\log\left(-\log(\widehat{S}(t))\right) = \log(\widehat{\lambda}) + \log(t) .$$

Hence,

$$\text{var}_a\left\{ \log\left(-\log(\widehat{S}(t))\right) \right\} = \text{var}_a\left(\log(\widehat{\lambda})\right) \approx \frac{1}{n_u} .$$

It follows from the delta method that for each fixed t, $\log(-\log(S(t))) = \log(\lambda t)$

$$\log\left(-\log(\widehat{S}(t))\right) \overset{a}{\sim} N\left(\log(\lambda t), \frac{1}{n_u}\right).$$

It then follows, with some algebraic manipulation, a $(1-\alpha)\times 100\%$ **C.I. for the true probability of survival beyond time t, $S(t)$,** is given by

$$\exp\left\{\log\left(\widehat{S}(t)\right)\exp\left(\frac{z_{\alpha/2}}{\sqrt{n_u}}\right)\right\} \leq S(t) \leq \exp\left\{\log\left(\widehat{S}(t)\right)\exp\left(\frac{-z_{\alpha/2}}{\sqrt{n_u}}\right)\right\}.$$

$$(3.20)$$

WHY!

- **The likelihood ratio test** (3.14):

$$
\begin{aligned}
r^*(\underline{y}) &= -2\log L(\lambda_0) + 2\log L(\widehat{\lambda}) \\
&= -2n_u\log(\lambda_0) + 2\lambda_0\sum_{i=1}^{n} y_i + 2n_u\log\left(\frac{n_u}{\sum_{i=1}^{n} y_i}\right) - 2n_u \\
&= -2\cdot 7\cdot\log(\frac{1}{30}) + \frac{2}{30}\cdot 423 + 2\cdot 7\cdot\log\left(\frac{7}{423}\right) - 2\cdot 7 \\
&= 4.396.
\end{aligned}
$$

The p-value $= P(r^*(\underline{y}) \geq 4.396) \approx 0.036$. Therefore, here we reject $H_0 : \theta = 1/\lambda = 30$ and conclude that mean survival is > 30 weeks.

A computer application:

We use the S function `survReg` to fit parametric models (with the MLE approach) for censored data. The following S program is intended to duplicate some of the previous hand calculations. It fits an exponential model to the AML data, yields point and 95% C.I. estimates for both the mean and the median, and provides a Q-Q plot for diagnostic purposes. Recall that the exponential model is just a Weibull with shape $\alpha = 1$ or, in log(time), is an extreme value model with scale $\sigma = 1$. The function `survReg` fits log(time) and outputs the coefficient $\hat{\mu} = -\log(\hat{\lambda})$, the MLE of μ, the location parameter of the extreme value distribution. Hence, the MLE(λ)=$\hat{\lambda} = \exp(-\hat{\mu})$ and the MLE(θ)= $\hat{\theta} = \exp(\hat{\mu})$. Unnecessary output has been deleted. The S function `predict` is a companion function to `survReg`. It provides estimates of quantiles along with their s.e.'s. One of the arguments of the `predict` function is `type`. Set `type="uquantile"` to produce estimates based on the log-transform as in Table 3.2. The default produces intervals based on the variance for quantiles derived on page 73. The function `qq.weibull` produces a Q-Q plot. The pound sign # denotes our inserted annotation. We store the data for the maintained group in a `data.frame` object called aml1. The two variables are `weeks` and `status`.

Exponential fit

```
> attach(aml1)
> exp.fit <- survReg(Surv(weeks,status)~1,dist="weib",scale=1)
> exp.fit
Coefficients:
 (Intercept)
    4.101457
Scale fixed at 1 Loglik(model)= -35.7 n= 11
```

The `Intercept` $= 4.1014$, which equals $\hat{\mu} = -\log(\hat{\lambda}) = \log(\hat{\theta})$. The next five line commands produce a 95% C.I. for the mean θ.

```
> coeff <- exp.fit$coeff # muhat
> var <- exp.fit$var
> thetahat <- exp(coeff) # exp(muhat)
> thetahat
     60.42828
> C.I.mean1 <- c(thetahat,exp(coeff-1.96*sqrt(var)),
              exp(coeff+1.96*sqrt(var)))
> names(C.I.mean1) <- c("mean1","LCL","UCL")
> C.I.mean1
```

```
       mean1        LCL          UCL
     60.42828    28.80787    126.7562
```

\# Estimated median along with a 95% C.I. (in weeks) using the predict function.

```
> medhat <- predict(exp.fit,type="uquantile",p=0.5,se.fit=T)
> medhat1 <- medhat$fit[1]
> medhat1.se <- medhat$se.fit[1]
> exp(medhat1)
[1] 41.88569
```

```
> C.I.median1 <- c(exp(medhat1),exp(medhat1-1.96*medhat1.se),
                   exp(medhat1+1.96*medhat1.se))
> names(C.I.median1) <- c("median1","LCL","UCL")
> C.I.median1
     median1        LCL          UCL
    41.88569    19.96809    87.86072
```

\# Point and 95% C.I. estimates for $S(t)$, the probability of survival beyond time t, at the uncensored maintained group's survival times.

```
> muhat <- exp.fit$coeff
> weeks.u <- weeks[status == 1]
> nu <- length(weeks.u)
> scalehat <- rep(exp(muhat),nu)
> Shat <- 1 - pweibull(weeks.u,1,scalehat)
  # In S, Weibull's scale argument is exp(muhat) = 1/lambdahat,
  # which we call scalehat.
> LCL <- exp(log(Shat)*exp(1.96/sqrt(nu)))#See expression (3.20)
> UCL <- exp(log(Shat)*exp(-1.96/sqrt(nu)))
> C.I.Shat <- data.frame(weeks.u,Shat,LCL,UCL)
> round(C.I.Shat,5)
     weeks.u    Shat       LCL       UCL   # 95% C.I.'s
1          9  0.86162  0.73168  0.93146
2         13  0.80644  0.63682  0.90253
4         18  0.74240  0.53535  0.86762
5         23  0.68344  0.45005  0.83406
7         31  0.59869  0.34092  0.78305
8         34  0.56970  0.30721  0.76473
10        48  0.45188  0.18896  0.68477
```

\# The next line command produces the Q-Q plot in Figure 3.8 using the qq.weibull function. The scale=1 argument forces an exponential to be fit.

```
> qq.weibull(Surv(weeks,status),scale=1)
[1] "qq.weibull:done"
```

The following table summarizes the estimates of the mean and the median.

Exponential fit with MLE to AML1 data		
	Point Estimate	95% C.I.
median1	41.88569	[19.968, 87.86] weeks
mean1	60.42828	[28.81, 126.76] weeks

This table's results match those in Table 3.2. In Figure 3.8 a Q-Q plot is displayed. The following S program performs a likelihood ratio test (LRT) of

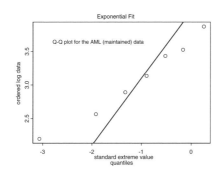

Figure 3.8 *Exponential Q-Q plot. The line has MLE intercept $\widehat{\mu}$ and slope 1.*

the null hypothesis $H_0 : \theta = 1/\lambda = 30$ weeks. To compute the value of the log likelihood function $L(\theta)$ at $\theta = 30$, we use the function `weib.loglik.theta`. It has four arguments: `time`, `status`, `shape`, `theta`. A shape value (α) of 1 forces it to fit an exponential and theta is set to $1/\lambda = 30$. The results match those hand-calculated back on page 74.

```
> weib.loglik.theta(weeks,status,1,30)
[1] -37.90838
> rstar <- - 2*(weib.loglik.theta(weeks,status,1,30) -
            exp.fit$loglik[1])
> rstar
[1] 4.396295
> pvalue <- 1 - pchisq(rstar,1)
> pvalue
[1] 0.0360171
```

3.4.2 Fitting data to the Weibull and log-logistic models

The following S program fits the AML data to the Weibull and log-logistic models both using the MLE approach via the `survReg` function. The `survReg`

function uses by default a log link function which transforms the problem into estimating location $\mu = -\log(\lambda)$ and scale $\sigma = 1/\alpha$. In the output from
> summary(weib.fit),

$\widehat{\mu}$ (= Intercept) <- weib.fit$coeff, and $\widehat{\sigma}$ (= Scale) <-weib.fit$scale.

This holds for any summary(fit) resulting from survReg evaluated at the "Weibull", "loglogistic", and "lognormal" distributions. The S output has been modified in that the extraneous output has been deleted.

Once the parameters are estimated via survReg, we can use S functions to compute estimated survival probabilities and quantiles. These functions are given in Table 3.3 for the reader's convenience.

Table 3.3:   *S distribution functions*

	Weibull	logistic $(Y = \log(T))$	normal $(Y = \log(T))$
$F(t)$	pweibull(q, α, λ^{-1})	plogis(q, μ, σ)	pnorm(q, μ, σ)
t_p	qweibull(p, α, λ^{-1})	qlogis(p, μ, σ)	qnorm(p, μ, σ)

Weibull fit

```
> weib.fit <- survReg(Surv(weeks,status)~1,dist="weib")
> summary(weib.fit)
              Value Std. Error      z        p
(Intercept)  4.0997      0.366  11.187 4.74e-029
 Log(scale) -0.0314      0.277  -0.113 9.10e-001
Scale= 0.969
```

Estimated median along with a 95% C.I. (in weeks).

```
> medhat <- predict(weib.fit,type="uquantile",p=0.5,se.fit=T)
> medhat1 <- medhat$fit[1]
> medhat1.se <- medhat$se.fit[1]
> exp(medhat1)
[1] 42.28842
> C.I.median1 <- c(exp(medhat1),exp(medhat1-1.96*medhat1.se),
                   exp(medhat1+1.96*medhat1.se))
> names(C.I.median1) <- c("median1","LCL","UCL")
> C.I.median1
   median1      LCL      UCL
  42.28842 20.22064 88.43986
> qq.weibull(Surv(weeks,status))   # Produces a Q-Q plot
[1] "qq.weibull:done"
```

Log-logistic fit

```
> loglogis.fit<-survReg(Surv(weeks,status)~1,dist="loglogistic")
> summary(loglogis.fit)
            Value Std. Error    z       p
(Intercept) 3.515    0.306   11.48 1.65e-030
Log(scale) -0.612    0.318   -1.93 5.39e-002
Scale= 0.542
```

Estimated median along with a 95% C.I. (in weeks).

```
> medhat <- predict(loglogis.fit,type="uquantile",p=0.5,se.fit=T)
> medhat1 <- medhat$fit[1]
> medhat1.se <- medhat$se.fit[1]
> exp(medhat1)
[1] 33.60127
> C.I.median1 <- c(exp(medhat1),exp(medhat1-1.96*medhat1.se),
                   exp(medhat1+1.96*medhat1.se))
> names(C.I.median1) <- c("median1","LCL","UCL")
> C.I.median1
    median1      LCL       UCL
   33.60127  18.44077  61.22549
> qq.loglogistic(Surv(weeks,status)) # Produces a Q-Q plot.
[1] "qq.loglogistic:done"
> detach()
```

Discussion

In order to compare some of the output readily, we provide a summary in the
following table:

MLE's fit to AML1 data at the models:

model	$\widehat{\mu}$	median1	95% C.I.	$\widehat{\sigma}$
exponential	4.1	41.88	[19.97, 87.86] weeks	1
Weibull	4.1	42.29	[20.22, 88.44] weeks	.969
log-logistic	3.52	33.60	[18.44, 61.23] weeks	.542

The log-logistic gives the narrowest C.I. among the three. Further, its estimated median of 33.60 weeks is the smallest and very close to the K-M
estimated median of 31 weeks on page 35. The Q-Q plots in Figure 3.10 are
useful for distributional assessment. It "appears" that a log-logistic model
fits adequately and is the best among the three distributions discussed. The
estimated log-logistic survival curve is overlayed on the K-M curve for the
AML1 data in Figure 3.9. We could also consider a log-normal model here.
The cautionary note, page 64, warns that we must compare the plot pattern
to the MLE line with slope $\widehat{\sigma}$ and y-intercept $\widehat{\mu}$. For without this comparison,

the least squares line alone fitted only to uncensored times would lead us to judge the Weibull survival model adequate. But, as we see in Figure 3.10, this is wrong. We do see that the least squares line in the Q-Q plot for the log-logistic fit is much closer to the MLE line with slope $\hat{\sigma}$ and y-intercept $\hat{\mu}$.

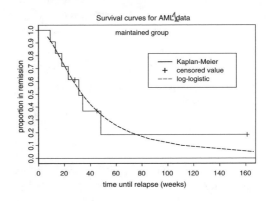

Figure 3.9 *K-M and log-logistic survival curves for AML data.*

3.5 Two-sample problem

In this section we compare two survival curves from the same parametric family. We focus on comparing the two scale (λ) parameters. In the log-transformed problem, this compares the two location, $\mu = -\log(\lambda)$, parameters. We picture this in Figure 3.11. We continue to work with the AML data. The nonparametric log-rank test (page 44) detected a significant difference (p-value= 0.03265) between the two K-M survival curves for the two groups, maintained and nonmaintained. We concluded maintenance chemotherapy prolongs remission period. We now explore if any of the log-transform distributions, which belong to the location and scale family (3.7), fit this data adequately. The **full model** can be expressed as a log-linear model as follows:

$$
\begin{aligned}
Y &= \log(T) \\
&= \tilde{\mu} + \text{error} \\
&= \theta + \beta^* \text{group} + \text{error} \\
&= \begin{cases} \theta + \beta^* + \text{error} & \text{if group} = 1 \text{ (maintained)} \\ \theta + \text{error} & \text{if group} = 0 \text{ (nonmaintained)}. \end{cases}
\end{aligned}
$$

The $\tilde{\mu}$ is called the *linear predictor*. In this two groups model, it has two values $\mu_1 = \theta + \beta^*$ and $\mu_2 = \theta$. Further, we know $\tilde{\mu} = -\log(\tilde{\lambda})$, where $\tilde{\lambda}$ denotes the scale parameter values of the distribution of the target variable T. Then $\tilde{\lambda} = \exp(-\theta - \beta^* \text{group})$. The two values are $\lambda_1 = \exp(-\theta - \beta^*)$ and

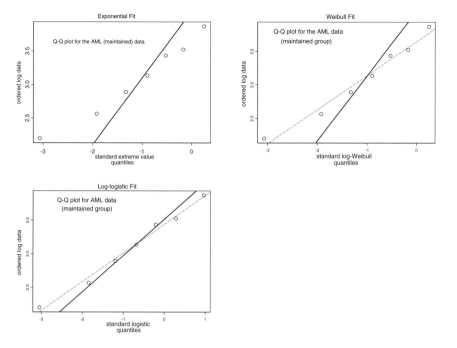

Figure 3.10 *Q-Q plots for the exponential, the Weibull, and the log-logistic. Each solid line is constructed with MLE's $\widehat{\mu}$ and $\widehat{\sigma}$. The dashed lines are least squares lines.*

$\lambda_2 = \exp(-\theta)$. The **null hypothesis** is:

$$H_0 : \lambda_1 = \lambda_2 \quad \text{if and only if} \quad \mu_1 = \mu_2 \quad \text{if and only if} \quad \beta^* = 0 \,.$$

Recall that the scale parameter in the log-transform model is the reciprocal of the shape parameter in the original model; that is, $\sigma = 1/\alpha$. We test H_0 under each of the following cases:

Case 1: Assume equal shapes (α); that is, we assume equal scales $\sigma_1 = \sigma_2 = \sigma$. Hence, error $= \sigma Z$, where the random variable Z has either the standard extreme value, standard logistic, or the standard normal distribution. Recall by standard, we mean $\mu = 0$ and $\sigma = 1$.

Case 2: Assume different shapes; that is, $\sigma_1 \neq \sigma_2$.

Fitting data to the Weibull, log-logistic, and log-normal models

In the following S program we first fit the AML data to the Weibull model and conduct formal tests. Then we fit the AML data to the log-logistic and log-normal models. Quantiles in the log-linear model setting are discussed. Lastly, we compare Q-Q plots. The S function `anova` conducts LRT's for hierarchical models; that is, each reduced model under consideration is a subset of the full

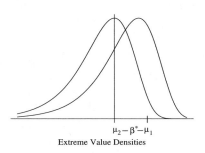

$$\mu_2 - \beta^* - \mu_1$$
Extreme Value Densities

Figure 3.11 *Comparison of two locations.*

model under consideration. Extraneous output has been deleted. The AML data is stored in the data frame aml.

Model 1: Data come from same distribution. **The Null Model** is $Y = \log(T) = \theta + \sigma Z$, where Z is a standard extreme value random variable.

```
> attach(aml)
> weib.fit0 <- survReg(Surv(weeks,status) ~ 1,dist="weib")
> summary(weib.fit0)
            Value Std. Error      z        p
(Intercept) 3.6425    0.217  16.780 3.43e-063 Scale= 0.912
Loglik(model)= -83.2   Loglik(intercept only)= -83.2
```

Model 2: Case 1: With different locations and equal scales σ, we express this model by

$$Y = \log(T) = \theta + \beta^* \text{group} + \sigma Z. \tag{3.21}$$

```
> weib.fit1 <- survReg(Surv(weeks,status) ~ group,dist="weib")
> summary(weib.fit1)
            Value Std. Error      z        p
(Intercept) 3.180     0.241  13.22 6.89e-040
      group 0.929     0.383   2.43 1.51e-002
Scale= 0.791 Loglik(model)= -80.5 Loglik(intercept only)= -83.2
    Chisq= 5.31 on 1 degrees of freedom, p= 0.021
> weib.fit1$linear.predictors # Extracts the estimated mutildes.
 4.1091 4.1091 4.1091 4.1091 4.1091 4.1091 4.1091 4.1091
 4.1091 4.1091 4.1091 3.1797 3.1797 3.1797 3.1797 3.1797
 3.1797 3.1797 3.1797 3.1797 3.1797 3.1797 3.1797
 # muhat1=4.109 and muhat2=3.18 for maintained and
 # nonmaintained groups respectively.
```

Model 3: Case 2: $Y = \log(T) = \theta + \beta^*$group + error, different locations, different scales.

Fit each group separately. On each group run a `survReg` to fit the data. This gives the MLE's of the two locations μ_1 and μ_2, and the two scales σ_1 and σ_2.

```
> weib.fit20 <- survReg(Surv(weeks,status) ~ 1,
                data=aml[aml$group==0,],dist="weib")
> weib.fit21 <- survReg(Surv(weeks,status) ~ 1,
                data=aml[aml$group==1,],dist="weib")
> summary(weib.fit20)
            Value  Std.Error     z         p
(Intercept) 3.222    0.198    16.25 2.31e-059  Scale=0.635
> summary(weib.fit21)
            Value  Std.Error     z         p
(Intercept) 4.1      0.366    11.19 4.74e-029  Scale=0.969
```

To test the reduced model against the full model we use the **LRT**. The anova function is appropriate for hierarchical models.

```
> anova(weib.fit0,weib.fit1,test="Chisq")
Analysis of Deviance Table Response: Surv(weeks, status)

  Terms Resid. Df    -2*LL Test Df  Deviance    Pr(Chi)
1     1       21   166.3573
2 group       20   161.0433     1  5.314048 0.02115415
# Model 2 is a significant improvement over the null
# model (Model 1).
```

To construct the appropriate likelihood function for Model 3 to be used in the LRT:

```
> loglik3 <- weib.fit20$loglik[2]+weib.fit21$loglik[2]
> loglik3
[1] -79.84817
> lrt23 <- -2*(weib.fit1$loglik[2]-loglik3)
> lrt23
[1] 1.346954
> 1 - pchisq(lrt23,1)
[1] 0.2458114  # Retain Model 2.
```

The following table summarizes the **three models weib.fit0, 1, and 2**:

Model	Calculated Parameters	The Picture
1 (0)	θ, σ	same location, same scale
2 (1)	$\theta, \beta^*, \sigma \equiv \mu_1, \mu_2, \sigma$	different locations, same scale
3 (2)	$\mu_1, \mu_2, \sigma_1, \sigma_2$	different locations, different scales

We now use the log-logistic and log-normal distribution to estimate Model 2. The form of the log-linear model is the same. The distribution of error terms is what changes.

$$Y = \log(T) = \theta + \beta^* \text{group} + \sigma Z,$$

where $Z \sim$ standard logistic or standard normal.

```
> loglogis.fit1 <- survReg(Surv(weeks,status) ~ group,
                           dist="loglogistic")
> summary(loglogis.fit1)
            Value Std. Error    z         p
(Intercept) 2.899     0.267  10.84 2.11e-027
      group 0.604     0.393   1.54 1.24e-001
Scale= 0.513 Loglik(model)= -79.4 Loglik(intercept only)= -80.6
Chisq= 2.41 on 1 degrees of freedom, p= 0.12 # p-value of LRT.
    # The LRT is test for overall model adequacy. It is not
    # significant.

> lognorm.fit1 <- survReg(Surv(weeks,status) ~ group,
                          dist="lognormal")
> summary(lognorm.fit1)
            Value Std. Error    z         p
(Intercept) 2.854     0.254  11.242 2.55e-029
      group 0.724     0.380   1.905 5.68e-002
Scale= 0.865 Loglik(model)= -78.9 Loglik(intercept only)= -80.7
Chisq= 3.49 on 1 degrees of freedom, p= 0.062 # p-value of LRT.
    # Here there is mild evidence of the model adequacy.
```

Quantiles

Let $\widehat{y}_p = \log(\widehat{t}_p)$ denote the estimated pth-quantile. For Model 2 (3.21) the quantile lines are given by

$$\widehat{y}_p = \widehat{\theta} + \widehat{\beta}^* \text{group} + \widehat{\sigma} z_p, \tag{3.22}$$

where z_p is the pth-quantile from either the standard normal, standard logistic, or standard extreme value tables. As p changes from 0 to 1, the standard quantiles z_p increase and $\widehat{y}_p$ is linearly related to z_p. The slope of the line is $\widehat{\sigma}$. There are two intercepts, $\widehat{\theta} + \widehat{\beta}^*$ and $\widehat{\theta}$, one for each group. Hence, we obtain two parallel quantile lines. Let us take z_p to be a standard normal quantile. Then if $p = .5$, $z_{.5} = 0$. Hence, $\widehat{y}_{.5} = \widehat{\theta} + \widehat{\beta}^* \text{group}$ represents the estimated median, and the mean as well, for each group. We see that if T is log-normal, then the estimated linear model $\widehat{y}_{.5} = \log(\widehat{t}_{.5}) = \widehat{\theta} + \widehat{\beta}^* \text{group}$ resembles the least squares line where we regress y to the group; that is, $\widehat{y} = \widehat{\theta} + \widehat{\beta}^* \text{group}$ is the estimated mean response for a given group. In Table 3.4 we provide the estimated $.10, .25, .50, .75, .90$ quantiles for the three error distributions under consideration. Plot any two points $(z_p, \widehat{y}_p)$ for a given group and distribution. Then draw a line through them. This is the MLE line drawn on the Q-Q plots in Figure 3.12.

The following S code computes point and C.I. estimates for the medians and draws Q-Q plots for the three different estimates of Model 2 (3.22). This recipe works for any desired estimated quantile. Just set p=desired quantile in the predict function.

Table 3.4: *Five quantiles for the AML data under Model 2* (3.22)

g	p	extreme value			logistic			normal		
		z_p	$\widehat{y}_p$	$\hat{t}_p$	z_p	$\widehat{y}_p$	$\hat{t}_p$	z_p	$\widehat{y}_p$	$\hat{t}_p$
	.10	-2.25	1.40	4.05	-2.20	1.77	5.88	-1.28	1.75	5.73
	.25	-1.25	2.19	8.98	-1.10	2.34	10.33	-.67	2.27	9.68
0	.50	-.37	2.89	18	0	2.9	18.16	0	2.85	17.36
	.75	.33	3.44	31.14	1.10	3.46	31.9	.67	3.44	31.12
	.90	.83	3.84	46.51	2.20	4.03	56.05	1.28	3.96	52.6
	.10	-2.25	2.33	10.27	-2.20	2.38	10.76	-1.28	2.47	11.82
	.25	-1.25	3.12	22.73	-1.10	2.94	18.91	-.67	2.99	20
1	.50	-.37	3.82	45.56	0	3.50	33.22	0	3.58	35.8
	.75	.33	4.37	78.84	1.10	4.07	58.36	.67	4.16	64.16
	.90	.83	4.77	117.77	2.2	4.63	102.53	1.28	4.69	108.5

g denotes group.

```
> medhat <- predict(weib.fit1,newdata=list(group=0:1),
                     type="uquantile",se.fit=T,p=0.5)
> medhat
$fit:
        1         2
   2.889819   3.81916
$se.fit:
   0.2525755 0.3083033
> medhat0 <- medhat$fit[1]
> medhat0.se <- medhat$se.fit[1]
> medhat1 <- medhat$fit[2]
> medhat1.se <- medhat$se.fit[2]

> C.I.median0 <- c(exp(medhat0),exp(medhat0-1.96*medhat0.se),
                   exp(medhat0+1.96*medhat0.se))
> names(C.I.median0) <- c("median0","LCL","UCL")
> C.I.median1 <- c(exp(medhat1),exp(medhat1-1.96*medhat1.se),
                   exp(medhat1+1.96*medhat1.se))
> names(C.I.median1) <- c("median1","LCL","UCL")
# Weibull 95% C.I.'s follow.
```

```
> C.I.median0
  median0      LCL       UCL
  17.99005  10.96568  29.51406
> C.I.median1
  median1      LCL       UCL
  45.56593  24.90045  83.38218
# Similarly, log-logistic 95% C.I.'s follow.
> C.I.median0
  median0      LCL       UCL
  18.14708  10.74736  30.64165
> C.I.median1
  median1      LCL       UCL
  33.21488  18.90175  58.36648
# Log-normal 95% C.I.'s follow.
> C.I.median0
  median0      LCL       UCL
  17.36382  10.55622  28.56158
> C.I.median1
  median1      LCL       UCL
  35.83274  20.50927  62.60512
# The Q-Q plots are next.
> t.s0 <- Surv(weeks[group==0],status[group==0])
> t.s1 <- Surv(weeks[group==1],status[group==1])
> qq.weibull(Surv(weeks,status))
> qq.weibreg(list(t.s0,t.s1),weib.fit1)
> qq.loglogisreg(list(t.s0,t.s1),loglogis.fit1)
> qq.lognormreg(list(t.s0,t.s1),lognorm.fit1)
> detach()
```

Results:

- The LRT per the anova function provides evidence that Model 2 (3.21), weib.fit1, which assumes equal scales, is adequate.

- We summarize the distributional fits to Model 2 (3.21) in the following table:

distribution	max. likeli $\log\left(L(\widehat{\theta},\widehat{\beta}^*)\right)$	p-value for model adequacy	$\widehat{\theta}$	$\widehat{\beta}^*$	p-value for group effect
Weibull	−80.5	0.021	3.180	0.929	0.0151
log-logistic	−79.4	0.12	2.899	0.604	0.124
log-normal	−78.9	0.062	2.854	0.724	0.0568

- For the Weibull fit we conclude that there is a significant "group" effect (p-value= 0.0151). The maintained group tends to stay in remission longer,

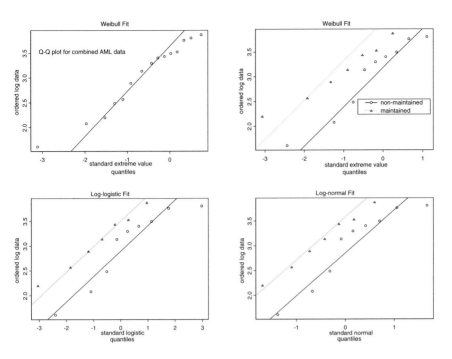

Figure 3.12 *Q-Q plots for the Weibull, the log-logistic, and the log-normal fit to Model 2:* $y = \theta + \beta^*\text{group} + \sigma Z$. *Each line constructed with the MLE's* $\widehat{\theta}, \widehat{\beta^*}$, *and* $\widehat{\sigma}$. *In each plot, the lines have same slope* $\widehat{\sigma}$ *and different intercepts, either* $\widehat{\theta}$ *or* $\widehat{\theta} + \widehat{\beta^*}$.

with estimated extreme value location parameters $\widehat{\mu}_1 = 4.109$ and $\widehat{\mu}_2 = 3.18$.

- The median of the maintained group is 45.6 weeks whereas the median of the nonmaintained group is only about 18 weeks. Corresponding 95% confidence intervals are (24.9, 83.38) weeks, and (10.96, 29.51) weeks, respectively.

- The log-normal has largest maximized likelihood, whereas the Weibull has the smallest. But the LRT for overall model fit is significant only for the Weibull; i.e., its p-value is the only one less than 0.05.

- The estimated linear predictor $\widehat{\widetilde{\mu}} = \widehat{\theta} + \widehat{\beta^*}\text{group}$. As $\widehat{\widetilde{\mu}} = -\log(\widehat{\widetilde{\lambda}}), \widehat{\widetilde{\lambda}} = \exp(-\widehat{\widetilde{\mu}}) = \exp(-\widehat{\theta} - \widehat{\beta^*}\text{group})$. $\widehat{\alpha} = 1/\widehat{\sigma}$. We summarize the estimated parameters for each group and distributional model in the following table:

	Weibull		log-logistic		log-normal	
group	$\widehat{\lambda}$	$\widehat{\alpha}$	$\widehat{\lambda}$	$\widehat{\alpha}$	$\widehat{\lambda}$	$\widehat{\alpha}$
0	0.042	1.264	0.055	1.95	0.058	1.16
1	0.0164	1.264	0.030	1.95	0.028	1.16

- The Q-Q plots in Figure 3.12 suggest that the log-logistic or log-normal models fit the maintained group data better than the Weibull model. However, they do not improve the fit for the nonmaintained.
- The nonparametric approach based on K-M, presented in Chapter 2, may give the better description of this data set.

Prelude to parametric regression models

As a prelude to parametric regression models presented in the next chapter, we continue to explore Model 2 (3.21) under the assumption that $T \sim$ Weibull. That is, we explore

$$
\begin{aligned}
Y &= \log(T) \\
&= \theta + \beta^* \text{group} + \sigma Z \\
&= \widetilde{\mu} + \sigma Z,
\end{aligned}
$$

where Z is a standard extreme minimum value random variable. Let the linear predictor $\widetilde{\mu} = -\log(\widetilde{\lambda})$ and $\sigma = 1/\alpha$. It follows from page 57 that the hazard function for the Weibull in this context is expressed as

$$
\begin{aligned}
h(t|\text{group}) &= \alpha \widetilde{\lambda}^{\alpha} t^{\alpha-1} \\
&= \alpha \lambda^{\alpha} t^{\alpha-1} \exp(\beta \text{group}) \\
&= h_0(t) \exp(\beta \text{group}), \quad\quad\quad (3.23)
\end{aligned}
$$

when we set $\lambda = \exp(-\theta)$ and $\beta = -\beta^*/\sigma$. WHY! The $h_0(t)$ denotes the baseline hazard; that is, when group $= 0$ or $\beta = 0$. Thus, $h_0(t)$ is the hazard function for the Weibull with scale parameter λ, which is free of any covariate.

The hazard ratio (HR) of group 1 to group 0 is

$$
\text{HR} = \frac{h(t|1)}{h(t|0)} = \frac{\exp(\beta)}{\exp(0)} = \exp(\beta).
$$

If we believe the Weibull model is appropriate, the HR is constant over time t. ~~The graph of $h(t|1)$ and $h(t|0)$ is a set of parallel lines over time.~~ We say the Weibull enjoys the *proportional hazards property* to be formally introduced in Chapter 4.3. On the AML data,

$$
\widehat{\beta} = \frac{-\widehat{\beta^*}}{\widehat{\sigma}} = \frac{-0.929}{0.791} = -1.1745.
$$

Therefore, the estimated HR is

$$
\widehat{\text{HR}} = \frac{\hat{h}(t|1)}{\hat{h}(t|0)} = \exp(-1.1745) \approx 0.31.
$$

The graph of HR
is horizontal line w/ height
$\exp(\beta)$.

The maintained group has 31% of the risk of the control group's risk of relapse. Or, the control group has $(1/0.31)=3.23$ times the risk of the maintained group of relapse at any given time t. The HR is a measure of effect that describes the relationship between time to relapse and group.

If we consider the ratio of the estimated survival probabilities, say at $t = 31$ weeks, since $\widehat{\lambda} = \exp(-\widehat{\mu})$, we obtain

$$\frac{\widehat{S}(31|1)}{\widehat{S}(31|0)} = \frac{0.652}{0.252} \approx 2.59 .$$

The maintained group is 2.59 times more likely to stay in remission at least 31 weeks. The Weibull survivor function $S(t)$ is given in a table on page 57.

3.6 A bivariate version of the delta method

$$\begin{pmatrix} x \\ y \end{pmatrix} \overset{a}{\sim} MVN \left(\begin{pmatrix} \mu_x \\ \mu_y \end{pmatrix} ; \begin{pmatrix} \sigma_x^2 & \sigma_{xy} \\ \sigma_{xy} & \sigma_y^2 \end{pmatrix} \right)$$

and suppose we want the asymptotic distribution of $g(x, y)$. Then the 1st order Taylor approximation for scalar fields is

$$g(x, y) \approx g(\mu_x, \mu_y) + (x - \mu_x)\frac{\partial}{\partial x}g(\mu_x, \mu_y) + (y - \mu_y)\frac{\partial}{\partial y}g(\mu_x, \mu_y).$$

Note that we expand about $(x, y) = (\mu_x, \mu_y)$. The $g(\cdot, \cdot)$ is a bivariate function that yields a scalar, i.e., a univariate. Then

$$g(x, y) \quad \overset{a}{\sim} \quad \text{normal with}$$
$$\text{mean} \approx g(\mu_x, \mu_y)$$
$$\text{asymptotic variance} \approx$$
$$\sigma_x^2(\tfrac{\partial}{\partial x}g)^2 + \sigma_y^2(\tfrac{\partial}{\partial y}g)^2 + 2\sigma_{xy}(\tfrac{\partial}{\partial x}g)(\tfrac{\partial}{\partial y}g). \quad (3.24)$$

WHY!

3.7 The delta method for a bivariate vector field

Let Σ denote the asymptotic covariance matrix of the random vector $(x, y)'$ given in Section 3.6. Suppose now we want the asymptotic distribution of the random vector

$$\underline{g}(x, y) = \begin{pmatrix} g_1(x, y) \\ g_2(x, y) \end{pmatrix}. \quad (3.25)$$

We need to employ a 1st order Taylor approximation for vector fields. This is summarized as follows: Expand $\underline{g}$ about $\underline{\mu} = (\mu_x, \mu_y)'$. Let A denote the

Jacobian matrix of $\underline{g}$ evaluated at $\underline{\mu}$. That is,

$$A \;=\; \begin{pmatrix} \frac{\partial g_1(\underline{\mu})}{\partial x} & \frac{\partial g_1(\underline{\mu})}{\partial y} \\[2mm] \frac{\partial g_2(\underline{\mu})}{\partial x} & \frac{\partial g_2(\underline{\mu})}{\partial y} \end{pmatrix}. \tag{3.26}$$

Then the 1st order Taylor approximation is

$$\underline{g}(x,y) \approx \begin{pmatrix} g_1(\mu_x,\mu_y) \\ g_2(\mu_x,\mu_y) \end{pmatrix} + A'\begin{pmatrix} x - \mu_x \\ y - \mu_y \end{pmatrix}. \tag{3.27}$$

The delta method now yields the following asymptotic distribution:

$$\underline{g}(x,y) \;\overset{a}{\sim}\; MVN \text{ with}$$

$$\text{mean vector} \approx \begin{pmatrix} g_1(\mu_x,\mu_y) \\ g_2(\mu_x,\mu_y) \end{pmatrix}$$

$$\text{covariance matrix} \approx A'\Sigma A. \tag{3.28}$$

Example: The covariance matrix of $(\widehat{\lambda}, \widehat{\alpha})'$

As noted in the Summary in Section 3.1, the log-transform of the Weibull, log-normal, and log-logistic models is a member of the location and scale family where $\mu = -\log(\lambda)$ and $\sigma = 1/\alpha$. The S function `survReg` only provides the estimated covariance matrix for $(\widehat{\mu}, \log(\widehat{\sigma}))'$. Exercises 3.12 and 3.13 challenge the reader to derive C.I.'s for $S(t)$ under the Weibull and log-logistic models, respectively. Then to actually compute the confidence bands for $S(t)$, we need to determine the asymptotic covariance matrix for $(\widehat{\lambda}, \widehat{\alpha})'$.

In these problems, $x = \widehat{\mu}$ and $y = \widehat{\sigma}^*$ where $\widehat{\sigma}^* = \log(\widehat{\sigma})$. So $\widehat{\lambda} = g_1(\widehat{\mu}) = \exp(-\widehat{\mu})$ and $\widehat{\alpha} = g_2(\widehat{\sigma}^*) = \exp(-\widehat{\sigma}^*)$. Expression (3.26) reduces to

$$A' = A = \begin{pmatrix} -\exp(-\mu) & 0 \\ 0 & -\exp(-\sigma^*) \end{pmatrix}$$

and expression (3.28) yields the covariance matrix

$$\text{cov}\begin{pmatrix} \widehat{\lambda} \\ \widehat{\alpha} \end{pmatrix} \approx \begin{pmatrix} \sigma_{\widehat{\mu}}^2 \exp(-2\mu) & \sigma_{\widehat{\mu},\widehat{\sigma}^*} \exp(-\mu - \sigma^*) \\ \sigma_{\widehat{\mu},\widehat{\sigma}^*} \exp(-\mu - \sigma^*) & \sigma_{\widehat{\sigma}^*}^2 \exp(-2\sigma^*) \end{pmatrix}. \tag{3.29}$$

The `survReg` function computes the estimated Σ, the covariance matrix of $\widehat{\mu}$ and $\widehat{\sigma}^* = \log(\widehat{\sigma})$. Use the following commands:

```
> fit <- survReg(....)
> summary(fit) # Outputs the estimates of mu and sigma star
> fit$var # Outputs the computed estimated covariance
          # matrix of these estimates.
```

Substitute the appropriate estimates into expression (3.29) to obtain the estimated covariance matrix of $(\widehat{\lambda}, \widehat{\alpha})'$.

3.8 General version of the likelihood ratio test

Let $X_1, X_2, \ldots, X_n$ denote a random sample from a population with p.d.f. $f(x|\theta)$, (θ may be a vector), where $\theta \in \Omega$, its parameter space. The **likelihood function** is given by

$$L(\theta) = L(\theta|\mathbf{x}) = \prod_{i=1}^{n} f(x_i|\theta), \text{ where } \mathbf{x} = (x_1, x_2, \ldots, x_n).$$

Let Ω_0 denote the null space. Then $\Omega = \Omega_0 \cup \Omega_0^c$.

Definition 3.8.1 *The likelihood ratio test statistic*

for testing $H_0 : \theta \in \Omega_0$ (reduced model) against $H_A : \theta \in \Omega_0^c$ (full model) is given by

$$r(\mathbf{x}) = \frac{\sup_{\Omega_0} L(\theta)}{\sup_{\Omega} L(\theta)}.$$

Note that $r(\mathbf{x}) \leq 1$. Furthermore, this handles hypotheses with nuisance parameters. Suppose $\theta = (\theta_1, \theta_2, \theta_3)$. We can test for example $H_0 : (\theta_1 = 0, \theta_2, \theta_3)$ against $H_A : (\theta_1 \neq 0, \theta_2, \theta_3)$. Here θ_2 and θ_3 are nuisance parameters. Most often, finding the sup amounts to finding the MLE's and then evaluating $L(\theta)$ at the MLE. Thus, for the denominator, obtain the MLE over the whole parameter space Ω. We refer to this as the full model. For the numerator, we maximize $L(\theta)$ over the reduced (restricted) space Ω_0. Find the MLE in Ω_0 and put into $L(\cdot)$. As $r(\mathbf{x}) \leq 1$, we reject H_0 for small values. Or, equivalently, we reject H_0 for large values of

$$r^*(\mathbf{x}) = -2 \log r(\mathbf{x}).$$

Theorem 3.8.1 *Asymptotic distribution of the $r^*(\mathbf{x})$ test statistic.*

Under $H_0 : \theta \in \Omega_0$, the distribution of $r^*(\mathbf{x})$ converges to a $\chi^2_{(df)}$ as $n \to \infty$. The degrees of freedom (df) = (# of free parameters in Ω) $-$ (# of free parameters $\in \Omega_0$).

That is,

$$r^*(\mathbf{x}) \overset{a}{\sim} \chi^2_{(df)}.$$

Proof: See Bickel & Doksum (2001, Chapter 6.3, Theorem 6.3.2).

Thus, an approximate size$-\alpha$ test is: reject H_0 iff $r^*(\mathbf{x}) = -2 \log r(\mathbf{x}) \geq \chi^2_\alpha$.

To compute approximate p-value: if $r^*(\mathbf{x}) = r^*$, then

$$p\text{-value} \approx P(r^*(\mathbf{x}) \geq r^*),$$

the area under the Chi-square curve to the right of r^*; that is, the upper tail area.

3.9 Exercises

A. *Applications*

3.1 Let T denote survival time of an experimental unit with survivor function $S(t) = \exp(-t/\theta)$ for $t > 0$ and $\theta > 0$. In this experiment n experimental units were observed and their lengths of life (in hours) were recorded. Let $t_1, \ldots, t_k$ denote the completely observed (uncensored) lifetimes, and let $c_{k+1}, c_{k+2}, \ldots, c_n$ denote the $n - k$ censored times. That is, this data set contains randomly right-censored data points.

(a) Derive the maximum likelihood estimate (MLE) $\widehat{\theta}_{ML}$ for θ. Describe this in words. Refer to expression (3.8) and pages 70– 71.

(b) Referring to the observed information matrix $i(\theta)$ (3.11), we derive the following expression for the (estimated) asymptotic variance of $\widehat{\theta}_{ML}$:

$$\widehat{\mathrm{var}}_a(\widehat{\theta}_{ML}) = \frac{(\widehat{\theta}_{ML})^2}{k},$$

where k is the number of uncensored data points n_u.

   i. Calculate for $t_1, \ldots, t_5 = 1, 4, 7, 9, 12$ and $c_6, c_7, \ldots, c_{10} = 3, 3, 4, 6, 6$ the value of the estimate $\widehat{\theta}_{ML}$ and $\widehat{\mathrm{var}}_a(\widehat{\theta}_{ML})$.

   ii. Provide an asymptotic 95% confidence interval for θ, the true mean lifetime of the experimental units. Refer to expression (3.18) and Table 3.2.

(c) Refer to expression (3.14). Give the expression for Neyman-Pearson/ Wilks Likelihood Ratio Test (LRT) statistic $r^*(t)$ to test the hypothesis $H_0 : \theta = \theta_0$. Then calculate its value on the data in part (b) with $\theta_0 = 10$. Use the asymptotic distribution of $r^*(t)$ to test the hypothesis $H_0 : \theta = 10$ against $H_A : \theta \neq 10$. Also see page 75.

(d) Suppose now that all n lifetimes are completely observed; that is, no censored times. Then $t_1, \ldots, t_5 = 1, 4, 7, 9, 12$ and $t_6, \ldots, t_{10} = 3, 3, 4, 6, 6$. Compute $\widehat{\theta}_{ML}$ and $\widehat{\mathrm{var}}_a(\widehat{\theta}_{ML})$. See page 68 and expression (3.11).

(e) For the complete data case in part (d), calculate the LRT statistic to test $H_0 : \theta = \theta_0$. Denote the LRT statistic by T_1^*. Use the asymptotic distribution of T_1^* to test $H_0 : \theta = 10$ against $H_A : \theta \neq 10$ with data from part (d). See page 71. Note that $T_1^* = r^*(\underline{t})$.

(f) In the complete data case there exists a test statistic T_2^*, equivalent to the LRT statistic T_1^*, to test $H_0 : \theta = \theta_0$:

$$T_2^* = \frac{2 \cdot \sum_{i=1}^{n} t_i}{\theta_0}.$$

The exact distribution of T_2^* under H_0 is $\chi^2_{(2n)}$.

   i. Conduct an analogous test to part (e) and compare results.

ii. Construct an exact 95% confidence interval for the true mean lifetime θ.

Hint: Refer to page 70.

3.2 We return to the diabetes data introduced in Exercise 2.3. We consider only the diabetic group. Refer to Section 3.4.2. In the output from
> summary(weib.fit), recall that $\hat{\mu}$ (= Intercept) <- weib.fit$coeff, and $\hat{\sigma}$ (= Scale) <- weib.fit$scale.

(a) Fit the data to the Weibull model. Then:

i. Obtain point and 95% C.I. estimates for the three quartiles.

ii. Compute point estimates for $S(t)$ at the uncensored survival times lzeit.

Tips:

```
> alphahat <- 1/weib.fit$scale
> lzeit.u <- sort(lzeit.u)
> Shat <- 1 - pweibull(lzeit.u,alphahat,scalehat)
```

iii. Plot the Kaplan-Meier curve and the estimated Weibull survivor function $(\hat{S}_W(t))$ Shat on the same plot.

Tips: Read the online help for par. Here it describes how to specify line types, etc.
```
> plot(....)
> lines(...., type="l", lty=2) # overlays the second
                               # graph on the same plot.
```

iv. Produce a Q-Q plot.

(b) Fit the data to the log-normal model. Repeat all of part (a).

Tips:

```
> lognorm.fit<-survReg(Surv(lzeit,tod)~1,dist="lognormal",
                       data=diabetes1)
> Shat <- 1 - pnorm(log(lzeit.u),lognorm.fit$coeff,
                    lognorm.fit$scale)
```

Plot Shat against lzeit.u (on the time axis). You must create your own qq.lognormal function. This is easy. Just read qq.loglogistic. Make minor changes.

(c) Compare your plots and comment as to how these models fit the data.

3.3 We continue to analyze the diabetes data.

(a) Investigate with the help of Q-Q plots which of the following Weibull models fit the lzeit data best: See summary table on page 83.

i. Model 1: Fit the diabetic and nondiabetic together without consideration of other variables.

   ii. Model 2: Fit the data with `diab` as a covariate in the model.

   iii. Model 3: In log(time), different locations and different scales.

Tips

For Models 1 and 3, use the function `qq.weibull`. For Model 2 use the function `qq.weibreg`. To plot all three on the same page, use

```
> par(mfrow=c(2,2))
> qq.weibull(....)
....
```

(b) Compare the log-likelihood values of these three models via the LRT statistics. The hierarchy is: Model 1 $\subset$ Model 2 $\subset$ Model 3.
Hint: See pages 82 and 83.

(c) Consider the fitted Model 2.

   i. Provide point estimates for the quartiles of the survival time (# days to death after operation) distribution of each `diab` group.

   ii. Provide point estimates of the two scale parameters and common shape parameter for the two Weibull survivor functions corresponding to the two `diab` groups.

   iii. Estimate the probability of surviving more than 1000 days.

B. *Theory and WHY!*

3.4 Derive the (estimated) asymptotic variance of the MLE $\widehat{\theta}_{ML}$

$$\widehat{\mathrm{var}}_a(\widehat{\theta}_{ML}) = \frac{(\widehat{\theta}_{ML})^2}{k},$$

which was stated back in Exercise 3.1(b).

3.5 Show that the LRT based on the statistic T_1^* in Exercise 3.1(e) is equivalent to the test based on the test statistic T_2^* presented in Exercise 3.1(f).
Hint: Show T_1^*, which is $r^*(\underline{t})$, is some convex function of T_2^*.

3.6 Prove the **Fact** stated on page 58. That is, if $T \sim$ Weibull, then $Y = \log(T) \sim$ extreme value distribution.

3.7 Derive the p.d.f. given in the table on page 59 of the log-normal r.v. T.

3.8 Show expression (3.15).
Hint: Refer to the **Example 6** in Hogg and Craig (1995, page 136).

3.9 Use expressions (3.18) and (3.19) to verify Table 3.2.

3.10 Derive expression (3.20).

3.11 Verify expression (3.23).

3.12 Attempt to derive a $(1 - \alpha)\%$ C.I. for $S(t)$ under the Weibull model.
Hint: See Sections 3.6 and 3.7.

3.13 Attempt to derive a $(1 - \alpha)\%$ C.I. for $S(t)$ under the log-logistic model.

3.14 Derive expression (3.24).
Hint: Use the first order Taylor approximation given above this expression.

## CHAPTER 4

Regression Models

Let T denote failure time and $\underline{x} = (x^{(1)}, \ldots, x^{(m)})'$ represent a vector of available **covariates**. We are interested in modelling and determining the relationship between T and $\underline{x}$. Often this is referred to as prognostic factor analysis. These $\underline{x}$ are also called regression variables, regressors, factors, or explanatory variables. The primary question is: Do any subsets of the m covariates help to explain survival time? For example, does age at first treatment and/or gender increase or decrease (relative) risk of survival? If so, how and by what estimated quantity?

Example 1. Let

- $x^{(1)}$ denote the sex ($x_i^{(1)} = 1$ for males and $x_i^{(1)} = 0$ for females),
- $x^{(2)} = $ Age at diagnosis,
- $x^{(3)} = x^{(1)} \cdot x^{(2)}$ (interaction),
- $T\dot{=}$survival time.

We introduce four models: the exponential, the Weibull, the Cox proportional hazards, and the accelerated failure time model, and a variable selection procedure.

Objectives of this chapter:

After studying Chapter 4, the student should:

1. Understand that the hazard function is modelled as a function of available covariates $\underline{x} = (x^{(1)}, \ldots, x^{(m)})'$.
2. Know that the **preferred link function** for $\eta = \underline{x}'\underline{\beta}$ is $\boldsymbol{k(\eta) = \exp(\eta)}$ and why.
3. Recognize the **exponential** and **Weibull regression models**.
4. Know the definition of the **Cox proportional hazards model**.
5. Know the definition of an **accelerated failure time model**.
6. Know how to compute the **AIC** statistic.
7. Know how to implement the S functions `survReg` and `predict` to estimate and analyze a parametric regression model and obtain estimated quantities of interest.

8. Know how to interpret the effects of a covariate on the risk and survivor functions.

4.1 Exponential regression model

We first generalize the exponential distribution. Recall that for the exponential distribution the hazard function, $h(t) = \lambda$, is constant with respect to time and that $E(T) = \frac{1}{\lambda}$. We model the hazard rate λ as a function of the covariate vector $\underline{x}$.

We assume the hazard function at time t for an individual has the form

$$h(t|\underline{x}) = h_0(t) \cdot k(\underline{x}'\underline{\beta}) = \lambda \cdot k(\underline{x}'\underline{\beta}) = \lambda \cdot k(\beta_1 x^{(1)} + \cdots + \beta_m x^{(m)}),$$

where $\beta = [\beta_1, \beta_2, \ldots, \beta_m]'$ is a vector of regression parameters (coefficients), $\lambda > 0$ is a constant, and k is a specified *link function*. The function $h_0(t)$ is called the baseline hazard. It's the value of the hazard function when the covariate vector $\underline{x} = \underline{0}$ or $\beta = \underline{0}$. Note that this hazard function is constant with respect to time t, but depends on $\underline{x}$.

The most natural choice for k is $k(x) = \exp(x)$, which implies

$$
\begin{aligned}
h(t|\underline{x}) &= \lambda \cdot \exp(\underline{x}'\underline{\beta}) \\
&= \lambda \cdot \exp\left(\beta_1 x^{(1)} + \cdots + \beta_m x^{(m)}\right) \\
&= \lambda \cdot \exp\left(\beta_1 x^{(1)}\right) \times \exp\left(\beta_2 x^{(2)}\right) \times \cdots \times \exp\left(\beta_m x^{(m)}\right).
\end{aligned}
$$

This says that the covariates act multiplicatively on the hazard rate. Equivalently, this specifies

$$\log(h(t|\underline{x})) = \log(\lambda) + \eta = \log(\lambda) + (\underline{x}'\underline{\beta}) = \log(\lambda) + \beta_1 x^{(1)} + \cdots + \beta_m x^{(m)}.$$

That is, the covariates act additively on the log failure rate – a log-linear model for the failure rate. The quantity $\eta = \underline{x}'\beta$ is called the *linear predictor of the log-hazard*. We may consider a couple of other k functions that may appear natural, $k(\eta) = 1 + \eta$ and $k(\eta) = 1/(1+\eta)$. The first one has a hazard function $h(t|\underline{x}) = \lambda \times (1 + \underline{x}'\beta)$ which is a linear function of $\underline{x}$ and the second has the mean $E(T|\underline{x}) = 1/h(t|\underline{x}) = (1 + \underline{x}'\beta)/\lambda$ which is a linear function of $\underline{x}$. Note that both proposals could produce **negative** values for the hazard (which is a violation) unless the set of β values is restricted to guarantee $k(\underline{x}'\beta) > 0$ for all possible $\underline{x}$. Therefore, $\boldsymbol{k(\eta) = \exp(\eta)}$ **is the most natural since it will always be positive no matter what the** $\underline{\beta}$ **and** $\underline{x}$ **are.**

The survivor function of T given $\underline{x}$ is

$$S(t|x) = \exp\left(-h(t|\underline{x})t\right) = \exp\left(-\lambda \exp(\underline{x}'\underline{\beta})t\right).$$

Thus, the p.d.f. of T given $\underline{x}$ is

$$f(t|\underline{x}) = h(t|\underline{x})S(t|\underline{x}) = \lambda \exp(\underline{x}'\underline{\beta})\exp\left(-\lambda \exp(\underline{x}'\underline{\beta})t\right).$$

Recall from **Fact**, Chapter 3.1, page 58, that if T is distributed exponentially, $Y = \log(T)$ is distributed as the extreme (minimum) value distribution with scale parameter $\sigma = 1$. Here, given $\underline{x}$, we have

$$\tilde{\mu} = -\log(h(t|\underline{x})) = -\log\left(\lambda \exp(\underline{x}'\underline{\beta})\right) = -\log(\lambda) - \underline{x}'\underline{\beta} \quad \text{and} \quad \sigma = 1.$$

Therefore, given $\underline{x}$,

$$Y = \log(T) = \tilde{\mu} + \sigma Z = \beta_0^* + \underline{x}'\underline{\beta}^* + Z,$$

where $\beta_0^* = -\log(\lambda)$, $\underline{\beta}^* = -\underline{\beta}$, and $Z \sim f(z) = \exp(z - e^z)$, $-\infty < z < \infty$, the standard extreme (minimum) value distribution.

In summary, $h(t|\underline{x}) = \lambda \exp(\underline{x}'\underline{\beta})$ is a log-linear model for the failure rate and transforms into a **linear** model for $Y = \log(T)$ in that the covariates act additively on Y.

Example 1 continued: The exponential distribution is usually a poor model for human survival times. We use it anyway for illustration. We obtain

hazard function:	$h(t|\underline{x}) = \lambda \exp(\underline{x}'\beta)$
log(hazard):	$\log(h(t|\underline{x})) = \log(\lambda) + \beta_1 x^{(1)} + \beta_2 x^{(2)} + \beta_3 x^{(3)}$
survivor function:	$S(t|\underline{x}) = \exp(-\lambda \exp(\underline{x}'\beta)t)$

	Male	Female
hazard	$\lambda \exp\left(\beta_1 + (\beta_2 + \beta_3)\text{age}\right)$	$\lambda \exp(\beta_2 \, \text{age})$
log(hazard)	$(\log(\lambda) + \beta_1) + (\beta_2 + \beta_3)\text{age}$	$\log(\lambda) + \beta_2 \, \text{age}$
survivor	$\exp\left(-\lambda \exp(\beta_1 + (\beta_2 + \beta_3)\text{age})t\right)$	$\exp\left(-\lambda \exp(\beta_2 \text{age})t\right)$

Take $\lambda = 1, \beta_1 = -1, \beta_2 = -0.2, \beta_3 = 0.1$. Then

	Male	Female
hazard	$\exp(-1 - .1 \cdot \text{age})$	$\exp(-0.2 \, \text{age})$
log(hazard)	$-1 - 0.1 \cdot \text{age}$	$-0.2 \cdot \text{age}$
survivor	$\exp\left(-\exp(-1 - 0.1 \cdot \text{age})t\right)$	$\exp\left(-\exp(-0.2 \cdot \text{age})t\right)$

Plots for this example are displayed in Figure 4.1.

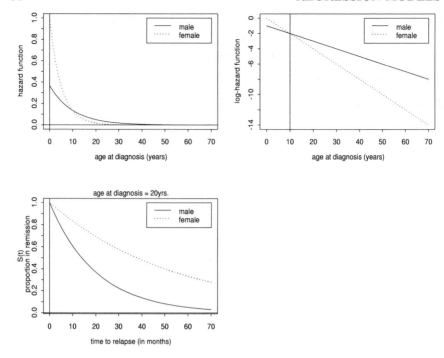

Figure 4.1 *Plots for Example 1.*

4.2 Weibull regression model

We generalize the Weibull distribution to regression in a similar fashion. Recall that its hazard function is $h(t) = \alpha \lambda^\alpha t^{\alpha-1}$.

To include the covariate vector $\underline{x}$ we now write the hazard for a given $\underline{x}$ as

$$
\begin{aligned}
h(t|\underline{x}) &= h_0(t) \cdot \exp(\underline{x}'\underline{\beta}) \qquad\qquad\qquad\qquad\qquad (4.1)\\
&= \alpha \lambda^\alpha t^{\alpha-1} \exp(\underline{x}'\underline{\beta}) = \alpha \left(\lambda \cdot \left(\exp(\underline{x}'\underline{\beta}) \right)^{\frac{1}{\alpha}} \right)^\alpha t^{\alpha-1}\\
&= \alpha (\tilde{\lambda})^\alpha t^{\alpha-1},
\end{aligned}
$$

where $\tilde{\lambda} = \lambda \cdot \left(\exp(\underline{x}'\underline{\beta}) \right)^{\frac{1}{\alpha}}$.

Again notice that

$$
\begin{aligned}
\log(h(t|\underline{x})) &= \log(\alpha) + \alpha \log(\tilde{\lambda}) + (\alpha - 1)\log(t)\\
&= \log(\alpha) + \alpha \log(\lambda) + \underline{x}'\underline{\beta} + (\alpha - 1)\log(t) .
\end{aligned}
$$

From **Fact**, Chapter 3.1, page 58, if $T \sim$ Weibull, then given $\underline{x}$, $Y = \log(T)$ $= \tilde{\mu} + \sigma Z$, where

$$
\tilde{\mu} = -\log(\tilde{\lambda}) = -\log(\lambda \cdot (\exp(\underline{x}'\underline{\beta}))^{\frac{1}{\alpha}}) = -\log(\lambda) - \frac{1}{\alpha}\underline{x}'\underline{\beta}, \qquad (4.2)
$$

$\sigma = \frac{1}{\alpha}$, and $Z \sim$ standard extreme value distribution. Therefore,

$$Y = \underbrace{\beta_0^* + \underline{x}'\underline{\beta}^*}_{\tilde{\mu}} + \sigma Z \,, \tag{4.3}$$

where $\beta_0^* = -\log(\lambda)$ and $\underline{\beta}^* = -\sigma\underline{\beta}$. It then follows from the table on page 57 that the survivor function of T given $\underline{x}$ is

$$S(t|\underline{x}) = \exp\left(-(\tilde{\lambda}t)^\alpha\right). \tag{4.4}$$

It follows from the relationship between the cumulative hazard and survivor functions given in expression (1.6) that, for a given $\underline{x}$, $H(t|\underline{x}) = -\log(S(t|\underline{x}))$. An expression for the log-cumulative hazard function follows from expression (4.2) for $\log(\tilde{\lambda})$.

$$\begin{aligned}
\log\left(H(t|\underline{x})\right) &= \alpha\log(\tilde{\lambda}) + \alpha\log(t) \\
&= \alpha\log(\lambda) + \alpha\log(t) + \underline{x}'\underline{\beta} \qquad (4.5) \\
&= \log\left(H_0(t)\right) + \underline{x}'\underline{\beta} \,,
\end{aligned}$$

where $H_0(t) = -\log\left(S_0(t)\right) = (\lambda t)^\alpha$. The log of the cumulative hazard function is linear in $\log(t)$ and in the β coefficients. Thus, for a fixed $\underline{x}$ value, the plot of $H(t|\underline{x})$ against t on a log-log scale is a straight line with slope α and intercept $\underline{x}'\underline{\beta} + \alpha\log(\lambda)$. Expression (4.5) can also be derived by noting expression (4.1) and definition (1.6) give

$$H(t|\underline{x}) = H_0(t)\exp(\underline{x}'\underline{\beta}) = (\lambda t)^\alpha \exp(\underline{x}'\underline{\beta}) \,. \tag{4.6}$$

In summary, for both the exponential and Weibull regression model, the effects of the covariates $\underline{x}$ act multiplicatively on the hazard function $h(t|\underline{x})$ which is clear from the form

$$\begin{aligned}
h(t|\underline{x}) &= h_0(t)\cdot\exp(\underline{x}'\underline{\beta}) = h_0(t)\cdot\exp\left(\beta_1 x^{(1)} + \cdots + \beta_m x^{(m)}\right) \\
&= h_0(t)\cdot\exp\left(\beta_1 x^{(1)}\right) \times \exp\left(\beta_2 x^{(2)}\right) \times \cdots \times \exp\left(\beta_m x^{(m)}\right).
\end{aligned}$$

This suggests the more general **Cox proportional hazards model**, presented in the next section. Further, both are log-linear models for T in that these models transform into a linear model for $Y = \log(T)$. That is, the covariates $\underline{x}$ act additively on $\log(T)$ (multiplicatively on T), which is clear from the form

$$Y = \log(T) = \tilde{\mu} + \sigma Z = \beta_0^* + \underline{x}'\underline{\beta}^* + \sigma Z \,.$$

This suggests a more general class of log-linear models called **accelerated failure time models** discussed in Section 4.4 of this chapter.

The difference from an ordinary linear regression model for the log-transformed target variable T, $Y = \log(T)$, is the distribution of the errors Z, which here is an extreme value distribution rather than a normal one. Therefore, least-squares methods are not adequate. Furthermore, there will be methods to deal with censored values, which is rarely discussed for ordinary linear regression.

4.3 Cox proportional hazards (PH) model

For the Cox PH model, the hazard function is

$$h(t|\underline{x}) = h_0(t) \cdot \exp(\underline{x}'\underline{\beta}), \tag{4.7}$$

where $h_0(t)$ is an unspecified baseline hazard function free of the covariates $\underline{x}$. The covariates act multiplicatively on the hazard. Clearly, the exponential and Weibull are special cases. At two different points $\underline{x}_1, \underline{x}_2$, the proportion

$$\frac{h(t|\underline{x}_1)}{h(t|\underline{x}_2)} = \frac{\exp(\underline{x}'_1\underline{\beta})}{\exp(\underline{x}'_2\underline{\beta})} = \exp\left((\underline{x}'_1 - \underline{x}'_2)\underline{\beta}\right), \tag{4.8}$$

called the **hazards ratio (HR)**, is constant with respect to time t. This defines the *proportional hazards property*.

Remark:

As with linear and logistic regression modelling, **a statistical goal of a survival analysis is to obtain some measure of effect that will describe the relationship between a predictor variable of interest and time to failure, after adjusting for the other variables we have identified in the study and included in the model.** In linear regression modelling, the measure of effect is usually the regression coefficient β. In logistic regression, the measure of effect is an odds ratio, the log of which is β for a change of 1 unit in x. **In survival analysis, the measure of effect is the hazards ratio (HR).** As is seen above, this ratio is also expressed in terms of an exponential of the regression coefficient in the model.

For example, let β_1 denote the coefficient of the group covariate with group $= 1$ if received treatment and group $= 0$ if received placebo. Put treatment group in the numerator of **HR**. A HR of 1 means that there is no effect. A hazards ratio of 10, on the other hand, means that the treatment group has ten times the hazard of the placebo group. Similarly, a HR of $1/10$ implies that the treatment group has one-tenth the hazard or risk of the placebo group.

Recall the relationship between hazard and survival is

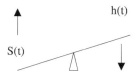

If the HR is less than one, then the ratio of corresponding survival probabilities is larger than one. Hence, the treatment group has larger probability of
survival at any given time t, after adjusting for the other covariates.

For any PH model, which includes the Weibull model as well as the Cox model,
the **survivor function** of T given $\underline{x}$ is

$$\boxed{S(t|\underline{x})} = \exp\left(-\int_0^t h(u|\underline{x})du\right) = \exp\left(-\exp(\underline{x}'\underline{\beta})\int_0^t h_0(u)du\right)$$

$$= \left(\exp\left(-\int_0^t h_0(u)du\right)\right)^{\exp(\underline{x}'\underline{\beta})} = \boxed{(S_0(t))^{\exp(\underline{x}'\underline{\beta})}},$$

where $S_0(t)$ denotes the baseline survivor function.

The p.d.f. of T given $\underline{x}$ is

$$f(t|\underline{x}) = h_0(t)\exp(\underline{x}'\underline{\beta})\,(S_0(t))^{\exp(\underline{x}'\underline{\beta})}\ .$$

There are two important generalizations:

(1) The baseline hazard $h_0(t)$ can be allowed to vary in specified subsets of the
data.
(2) The regression variables $\underline{x}$ can be allowed to depend on time; that is, $\underline{x} = \underline{x}(t)$.

We devote Chapter 5 of this book to an example of a Cox PH prognostic
factor analysis. Using S functions we analyze the CNS data. In Chapter 7.1 we
present an example which violates the PH assumption. The example explores
an epidemiologic study on the treatment of heroin addicts. To model and
compare the retention times of two clinics that differ strongly in their overall
treatment policies were the primary goals of the study. The PH assumption
is violated for the primary exposure variable of interest, clinic. An **extended
Cox model** is implemented to accommodate this kind of time dependency.
The Cox model is compared with an alternative regression quantile analysis
in Chapter 8.

4.4 Accelerated failure time model

This model is a log-linear regression model for T in that we model $Y = \log(T)$
as a linear function of the covariate $\underline{x}$. Suppose

$$Y = \underline{x}'\underline{\beta}^* + Z^*,$$

where Z^* has a certain distribution. Then

$$T = \exp(Y) = \exp(\underline{x}'\underline{\beta}^*) \cdot \exp(Z^*) = \exp(\underline{x}'\underline{\beta}^*) \cdot T^*,$$

where $T^* = \exp(Z^*)$. Here the covariate $\underline{x}$ acts multiplicatively on the survival time T. Suppose further that T^* has hazard function $h_0^*(t^*)$ which is independent of $\underline{\beta}^*$; that is, free of the covariate vector $\underline{x}$. The hazard function of T for a given $\underline{x}$ can be written in terms of the baseline function $h_0^*(\cdot)$ according to

$$h(t|\underline{x}) = h_0^*(\exp(-\underline{x}'\underline{\beta}^*)t) \cdot \exp(-\underline{x}'\underline{\beta}^*). \tag{4.9}$$

We prove this at the end of this section. We see here that the covariates $\underline{x}$ act multiplicatively on both t and the hazard function. The log-logistic and log-normal regression models are examples of accelerated failure time models as well as the exponential and Weibull regression models.

To illustrate this, let's consider the log-logistic model. As is customary, we first write this as a log-linear model

$$Y = \beta_0^* + \underline{x}'\underline{\beta}^* + \sigma Z,$$

where $Z \sim$ standard logistic. Let $Z^* = \beta_0^* + \sigma Z$. Then this log-linear model is now expressed as

$$Y = \underline{x}'\underline{\beta}^* + Z^*,$$

where $Z^* \sim$ logistic with mean being β_0^* and scale parameter σ. According to the table on page 60, the baseline hazard function for the log-logistic time T^* is

$$h_0^*(t^*) = \frac{\alpha\lambda^\alpha(t^*)^{\alpha-1}}{1 + \lambda^\alpha(t^*)^\alpha},$$

where $\beta_0^* = -\log(\lambda)$ and $\sigma = \alpha^{-1}$. This baseline hazard is free of β^*. Hence, this log-linear model is indeed an accelerated failure time model. It follows directly from expression (4.9) that the hazard function for the target random variable T is given by

$$h(t|\underline{x}) = \frac{\alpha\tilde{\lambda}^\alpha t^{\alpha-1}}{1 + \tilde{\lambda}^\alpha t^\alpha}, \tag{4.10}$$

where $\tilde{\lambda} = \lambda \cdot \left(\exp(-\underline{x}'\underline{\beta}^*)\right)$. WHY! Now, starting from here, as the hazard function uniquely defines a distribution, it follows from page 60 that for a given $\underline{x}$, $T \sim$ log-logistic with parameters $\tilde{\lambda}$ and α. Thus,

$$Y = \log(T) = \tilde{\mu} + \sigma Z, \tag{4.11}$$

where

$$\tilde{\mu} = -\log(\tilde{\lambda}) = -\log(\lambda) + \underline{x}'\underline{\beta}^* = \beta_0^* + \underline{x}'\underline{\beta}^*,$$

and $Z \sim$ standard logistic. It is important to note that the form of the above hazard function shows us the log-logistic model, although it is a log-linear model, is not a PH model defined in the previous section of this chapter.

It follows from expressions (1.6) and (4.9) that the **survivor function** of T

given $\underline{x}$ is

$$S(t|\underline{x}) = \exp\left(-\exp(-\underline{x}'\underline{\beta}^*)\int_0^t h_0^*\left(\exp(-\underline{x}'\underline{\beta}^*)u\right)du\right). \qquad (4.12)$$

Change the integration variable to $v = \exp(-\underline{x}'\underline{\beta}^*)u$. Then $dv = \exp(-\underline{x}'\underline{\beta}^*)du$ and $0 < v < \exp(-\underline{x}'\underline{\beta}^*)t$. Then for the accelerated failure time model,

$$\boxed{S(t|\underline{x})} = \exp\left(-\int_0^{\exp(-\underline{x}'\underline{\beta}^*)t} h_0^*(v)dv\right) = \boxed{S_0^*\left(\exp(-\underline{x}'\underline{\beta}^*)t\right)} = S_0^*(t^*),$$
$$(4.13)$$

where $S_0^*(t)$ denotes the baseline survivor function. Here we notice that the role of the covariate $\underline{x}$ changes the scale of the horizontal (t) axis. For example, if $\underline{x}'\underline{\beta}^*$ increases, then the last term in expression (4.13) increases. In this case it has decelerated the time to failure. This is why the log-linear model defined here is called the accelerated (decelerated) failure time model.

Remarks:

1. We have seen that the Weibull regression model, which includes the exponential, is a special case of both the Cox PH model and the accelerated failure time model. It is shown on pages 34 and 35 of Kalbfleisch and Prentice (1980) that the only log-linear models that are also PH models are the Weibull regression models.

2. Through the **partial likelihood** (Cox, 1975) we obtain estimates of the coefficients β that require no restriction on the baseline hazard $h_0(t)$. The S function coxph implements this. This partial likelihood is heuristically derived in Chapter 6.

3. For the accelerated failure time models we specify the baseline hazard function $h_0(t)$ by specifying the distribution function of Z^*.

4. Proof of (4.9): Recall a general result treated in an introductory mathematical statistics course. Let T^* have a p.d.f. $f^*(\cdot)$. Suppose we define a new random variable $T = g(T^*)$. Let $f(t)$ denote the p.d.f. of T. Then

$$f(t) = f^*\left(g^{-1}(t)\right) \cdot \left|\frac{dg^{-1}(t)}{dt}\right|.$$

Let $S(t)$ denote the survivor function of T. Then it follows that $\;$ if $g(.)$

is an increasing function
$$S(t) = S^*\left(g^{-1}(t)\right), \qquad (4.14)$$

where $S^*(\cdot)$ is the survivor function of T^*. WHY! Let T^* have a hazard function $h_0^*(\cdot)$. This implies that T^* has survivor function $S^*(t) = \exp\left(-\int_0^t h_0^*(u)du\right)$ and p.d.f. $f^*(t) = h_0^*(t)\exp\left(-\int_0^t h_0^*(u)du\right)$. Now, $t = g(t^*) = \exp(\underline{x}'\underline{\beta}^*)t^*$ and so $t^* = g^{-1}(t) = \exp(-\underline{x}'\underline{\beta}^*)t$ and $\frac{d}{dt}(g^{-1}(t)) =$

$\exp(-\underline{x}'\underline{\beta}^*)$. Then, in general, the hazard function for T is given by

$$h(t) = \frac{f(t)}{S(t)} = \frac{f^*\left(g^{-1}(t)\right) \cdot \left|\frac{dg^{-1}(t)}{dt}\right|}{S^*\left(g^{-1}(t)\right)} = h_0^*\left(g^{-1}(t)\right) \cdot \left|\frac{dg^{-1}(t)}{dt}\right|$$

by definition of $h_0^*(\cdot)$. To show its dependence on the covariate $\underline{x}$, we write this as

$$h(t|\underline{x}) = h_0^*\left(\exp(-\underline{x}'\underline{\beta}^*)t\right) \cdot \exp(-\underline{x}'\underline{\beta}^*) \ .$$

5. Hosmer and Lameshow (1999) well present the *proportional odds and proportional times properties* of the log-logistic regression model. From expression (4.13) and page 60 we can express the log-logistic's survivor function as

$$S(t|\underline{x}, \beta_0^*, \underline{\beta}^*, \alpha) = \frac{1}{1 + \exp(\alpha(y - \beta_0^* - \underline{x}'\underline{\beta}^*))} \ , \tag{4.15}$$

where $y = \log(t)$, $\beta_0^* = -\log(\lambda)$, and $\alpha = 1/\sigma$. WHY! The odds of survival beyond time t is given by

$$\frac{S(t|\underline{x}, \beta_0^*, \underline{\beta}^*, \alpha)}{1 - S(t|\underline{x}, \beta_0^*, \underline{\beta}^*, \alpha)} = \exp(-\alpha(y - \beta_0^* - \underline{x}'\underline{\beta}^*)). \tag{4.16}$$

Note that $-\log(\text{odds})$ is both a linear function of $\log(t)$ and the covariates $x^{(j)}$'s, $j = 1, \ldots, m$. The odds-ratio of survival beyond time t evaluated at $\underline{x}_1$ and $\underline{x}_2$ is given by

$$\text{OR}(t|\underline{x} = \underline{x}_2, \underline{x} = \underline{x}_1) = \exp(\alpha(\underline{x}_2 - \underline{x}_1)'\underline{\beta}^*). \tag{4.17}$$

The odds-ratio is commonly used as a measure of the effects of covariates. Note that the ratio is independent of time, which is referred to as the *proportional odds property*. For example, if OR $= 2$, then the odds of survival beyond time t among subjects with $\underline{x}_2$ is twice that of subjects with $\underline{x}_1$, and this holds for all t. Alternatively, some researchers prefer to describe the effects of covariates in terms of the survival time. The $(p \times 100)$th percentile of the survival distribution is given by

$$t_p(\underline{x}, \beta_0^*, \underline{\beta}^*, \alpha) = \left(p/(1-p)\right)^\sigma \exp(\beta_0^* + \underline{x}'\underline{\beta}^*). \tag{4.18}$$

WHY! Then, for example, the times-ratio at the median is

$$\text{TR}(t_{.5}|\underline{x} = \underline{x}_2, \underline{x} = \underline{x}_1) = \exp((\underline{x}_2 - \underline{x}_1)'\underline{\beta}^*). \tag{4.19}$$

This holds for any p. The TR is constant with respect to time, which is referred to as the *proportional times property*. Similarly, if TR $= 2$, then the survival time among subjects with $\underline{x}_2$ is twice that of subjects with $\underline{x}_1$, and this holds for all t. The upshot is that OR $= \text{TR}^\alpha$. That is, the odds-ratio is the power of the time ratio. Hence, the rate of change of OR is controlled by α, the shape parameter of the log-logistic distribution. For $\alpha = 1$, OR $=$ TR. If $\alpha = 2$ and TR $= 2$, then OR $= 2^2 = 4$. For one unit increase in a single component, fixing the other components in $\underline{x}$,

OR $\to +\infty$ or 0 as $\alpha \to \infty$ depending on the sign of the corresponding component of $\underline{\beta}^*$, and $\to 1$ as $\alpha \to 0$. Finally, Cox and Oakes (1984, page 79) claim that the log-logistic model is the only accelerated failure time model with the *proportional odds property*; equivalently, the only model with the *proportional times property*.

4.5 Summary

Let Z denote either a standard extreme value, standard logistic, or standard normal random variable. That is, each has location $\mu = 0$ and scale $\sigma = 1$.

-

$$Y = \log(T) = \widetilde{\mu} + \sigma Z = \beta_0^* + \underline{x}'\underline{\beta}^* + \sigma Z$$

accelerated failure time model
log-linear model

$\nearrow \swarrow$	$\uparrow\downarrow$	$\searrow\nwarrow$
T	T	T
Weibull	log-logistic	log-normal
$\downarrow$	$\downarrow$	
PH property	proportional odds property	
	$\updownarrow$	
	proportional times property	

The $\widetilde{\mu}$ is called the *linear predictor* and σ is the *scale parameter*. In the target variable T distribution, $\widetilde{\lambda} = \exp(-\widetilde{\mu})$ and the shape $\alpha = 1/\sigma$. The S function `survReg` estimates β_0^*, $\underline{\beta}^*$, and σ. The `predict` function provides estimates of $\widetilde{\mu}$ at specified values of the covariates. For example, returning to the AML data, where we have one covariate "group" with two values 0 or 1, to estimate the linear predictor (`lp`) for the maintained group, use
> `predict(fit,type="lp",newdata=list(group=1))`.

- The Weibull regression model is the only log-linear model that has the proportional hazards property. For both the Cox PH model and the Weibull regression model, we model the hazard function

$$h(t|\underline{x}) = h_0(t) \cdot \exp(\underline{x}'\underline{\beta}),$$

where $h_0(t)$ is the baseline hazard function. For the Weibull model, the baseline hazard $h_0(t) = \alpha\lambda^\alpha t^{\alpha-1}$, the baseline cumulative hazard $H_0(t) = (\lambda t)^\alpha$, and the log-cumulative hazard

$$\log\left(H(t|\underline{x})\right) = \alpha\log(\lambda) + \alpha\log(t) + \underline{x}'\underline{\beta}.$$

For the Weibull model, the relationship between the coefficients in the log-linear model and coefficients in modelling the hazard function is

$$\underline{\beta} = -\sigma^{-1}\underline{\beta}^* \quad \text{and} \quad \lambda = \exp(-\beta_0^*) \, .$$

The S function `survReg` estimates β_0^*, $\underline{\beta}^*$, and σ. The hazard ratio is

$$\text{HR}(t|\underline{x} = \underline{x}_2, x = \underline{x}_1) = \frac{h(t|\underline{x}_2)}{h(t|\underline{x}_1)} = \left(\exp\left((\underline{x}_1' - \underline{x}_2')\underline{\beta}^* \right) \right)^{\frac{1}{\sigma}} \, .$$

Fitting data to a Cox PH model is presented in detail in Chapter 5. The Cox procedure estimates the $\underline{\beta}$ coefficients directly.

- The log-logistic regression model is the only log-linear model that has the proportional odds property. The survivor function is

$$S(t|\underline{x}) = S_0^* \left(\exp(-\underline{x}'\underline{\beta}^*) \right) = \frac{1}{1 + \left(\exp(y - \beta_0^* - \underline{x}'\underline{\beta}^*) \right)^{\frac{1}{\sigma}}} \, ,$$

where $S_0^*(t)$ is the baseline survivor function, $y = \log(t)$, $\beta_0^* = -\log(\lambda)$, and $\alpha = 1/\sigma$.

The odds of survival beyond time t is given by

$$\frac{S(t|\underline{x})}{1 - S(t|\underline{x})} = \left(\exp(y - \beta_0^* - \underline{x}'\underline{\beta}^*) \right)^{-\frac{1}{\sigma}} \, .$$

The $(p \times 100)$th percentile of the survival distribution is given by

$$t_p(\underline{x}) = \left(p/(1-p) \right)^{\sigma} \exp(\beta_0^* + \underline{x}'\underline{\beta}^*).$$

The odds-ratio of survival beyond time t evaluated at $\underline{x}_1$ and $\underline{x}_2$ is given by

$$\text{OR}(t|\underline{x} = \underline{x}_2, \underline{x} = \underline{x}_1) = \left(\exp\left((\underline{x}_2 - \underline{x}_1)'\underline{\beta}^* \right) \right)^{\frac{1}{\sigma}} = \left(\text{TR} \right)^{\frac{1}{\sigma}} \, ,$$

where TR is the times-ratio. The reciprocal of the OR has the same functional form as the HR in the Weibull model with respect to $\underline{\beta}^*$ and σ.

- The upshot is: to obtain the estimated measures of effect, $\widehat{\text{HR}}$ and $\widehat{\text{OR}}$, we need only the estimates given by `survReg`.

4.6 AIC procedure for variable selection

Akaike's information criterion (AIC):

Comparisons between a number of possible models, which need not necessarily be nested nor have the same error distribution, can be made on the basis of the statistic

$$\text{AIC} = -2 \times \log(\text{maximum likelihood}) + k \times p,$$

where p is the number of parameters in each model under consideration and k

a predetermined constant. This statistic is called **Akaike's (1974) informa-tion criterion (AIC)**; the smaller the value of this statistic, the better the model. This statistic trades off goodness of fit (measured by the maximized log likelihood) against model complexity (measured by p). Here we shall take k as 2. For other choice of values for k, see the remarks at the end of this section.

We can rewrite the AIC to address parametric regression models considered in the text. For the parametric models discussed, the AIC is given by

$$AIC = -2 \times \log(\text{maximum likelihood}) + 2 \times (a + b), \qquad (4.20)$$

where a is the number of parameters in the specific model and b the number of one-dimensional covariates. For example, $a = 1$ for the exponential model, $a = 2$ for the Weibull, log-logistic, and log-normal models.

Here we manually step through a sequence of models as there is only one one-dimensional covariate. But in Chapter 5 we apply an automated model selec-tion procedure via an S function `stepAIC` as there are many one-dimensional covariates.

Motorette data example:

The data set given in Table 4.1 below was obtained by Nelson and Hahn (1972) and discussed again in Kalbfleisch and Prentice (1980), on pages 4, 5, 58, and 59. Hours to failure of motorettes are given as a function of operating tem-peratures 150^0C, 170^0C, 190^0C, or 220^0C. There is severe (Type I) censoring, with only 17 out of 40 motorettes failing. Note that the stress (temperature) is constant for any particular motorette over time. The primary purpose of the experiment was to estimate certain percentiles of the failure time distribution at a design temperature of 130^0C. We see that this is an accelerated process. The experiment is conducted at higher temperatures to speed up failure time. Then they make predictions at a lower temperature that would have taken them much longer to observe. The authors use the single regressor variable $x = 1000/(273.2+\text{Temperature})$. They also omit all ten data points at tem-perature level of 150^0C. We also do this in order to compare our results with Nelson and Hahn and Kalbfleisch and Prentice. We entered the data into a data frame called **motorette**. It contains

time	status	temp	x
hours	1 if uncensored 0 if censored	^{0}C	$1000/(273.2+{}^0$C$)$

We now fit the exponential, Weibull, log-logistic, and log-normal models. The log likelihood and the AIC for each model are reported in Table 4.2. The S

Table 4.1: *Hours to failure of Motorettes*

Temperature	Times
150°C	All 10 motorettes without failure at 8064 hours
170°C	1764, 2772, 3444, 3542, 3780, 4860, 5196
	3 motorettes without failure at 5448 hours
190°C	408, 408, 1344, 1344, 1440
	5 motorettes without failure at 1680 hours
220°C	408, 408, 504, 504, 504
	5 motorettes without failure at 528 hours
$n = 40$,	n_u = no. of uncensored times = 17

Table 4.2: *Results of fitting parametric models to the Motorette data*

Model		log-likelihood	AIC		
exponential	intercept only	-155.875	311.750 + 2(1)	= 313.750
	both	-151.803	303.606 + 2(1 + 1)	= 307.606
Weibull	intercept only	-155.681	311.363 + 2(2)	= 315.363
	both	-144.345	288.690 + 2(2 + 1)	= 294.690
log-logistic	intercept only	-155.732	311.464 + 2(2)	= 315.464
	both	-144.838	289.676 + 2(2 + 1)	= 295.676
log-normal	intercept only	-155.018	310.036 + 2(2)	= 314.036
	both	-145.867	291.735 + 2(2 + 1)	= 297.735

code for computing the AIC follows next. For each of these models the form
is the same:

$$\text{intercept only:} \quad y = \log(\hat{t}) = \beta_0^* + \sigma Z$$
$$\text{both:} \quad y = \log(\hat{t}) = \beta_0^* + \beta_1^* + \sigma Z,$$

where the distributions of Z are standard extreme (minimum) value, standard
logistic, and standard normal, respectively.

The S code for computing the AIC for a number of specified distributions

```
> attach(motorette) # attach the data frame motorette to avoid
                     # continually referring to it.
 # Weibull fit
> weib.fit <- survReg(Surv(time,status)~x,dist="weibull")
> weib.fit$loglik # the first component for intercept only and
                  # the second for both
[1] -155.6817 -144.3449
> -2*weib.fit$loglik # -2 times maximum log-likelihood
```

```
[1] 311.3634 288.6898
 # exponential fit
> exp.fit <- survReg(Surv(time,status)~x,dist="exp")
> -2*exp.fit$loglik
[1] 311.7501 303.6064
 # log-normal fit
> lognormal.fit <- survReg(Surv(time,status)~x,
                           dist="lognormal")
> -2*lognormal.fit$loglik
[1] 310.0359 291.7345
 # log-logistic fit
> loglogistic.fit <- survReg(Surv(time,status)~x,
                             dist="loglogistic")
> -2*loglogistic.fit$loglik
[1] 311.4636 289.6762
> detach() # Use this to detach the data frame when no
           # longer in use.
```

Nelson and Hahn applied a log-normal model, and Kalbfleisch and Prentice applied a Weibull model. Kalbfleisch and Prentice state that the Weibull model is to some extent preferable to the log-normal on account of the larger maximized log likelihood. From Table 4.2, we find that the Weibull distribution provides the best fit to this data, the log-logistic distribution is a close second, and the log-normal distribution is the third.

When there are no subject matter grounds for model choice, the model chosen for initial consideration from a set of alternatives might be the one for which the value of AIC is a minimum. It will then be important to confirm that the model does fit the data using the methods for model checking described in Chapter 6. We revisit AIC in the context of the PH regression model in Chapter 5.

Remarks:

1. In his paper (1974), Akaike motivates the need to develop a new model identification procedure by showing the standard hypothesis testing procedure is not adequately defined as a procedure for statistical model identification. He then introduces AIC as an appropriate procedure of statistical model identification.

2. Choice of k in the AIC seems to be flexible. Collett (1994) states that the choice $k = 3$ in the AIC is roughly equivalent to using a 5% significance level in judging the difference between the values of $-2 \times \log(\text{maximum likelihood})$ for two nested models which differ by one to three parameters. He recommends $k = 3$ for general use.

3. There are a variety of model selection indices similar in spirit to AIC. These are, going by name, BIC, Mallow's C_p, adjusted R^2, $R_a^2 = 1 -$

$(1 - R^2)(n - 1)/(n - p)$, where p is the number of parameters in the least squares regression, and some others. These all adjust the goodness of fit of the model by penalizing for complexity of the model in terms of the number of parameters.

4. Efron (1998) cautions that the validity of the selected model through currently available methods may be doubtful in certain situations. He illustrates an example where a bootstrap simulation study certainly discourages confidence in the selected model. He and his student find that from 500 bootstrap sets of data there is only one match to the originally selected model. Further, only one variable in the originally selected model appears in more than half (295) of the bootstrap set based models.

5. Bottom line in model selection: Does it make sense!

Estimation and testing: fitting the Weibull model

The S function `survReg` fits the times T as log-failure times $Y = \log(T)$ to model (4.3)

$$Y = \beta_0^* + \underline{x}'\underline{\beta}^* + \sigma Z,$$

where Z has the standard extreme value distribution. Further, when we re-express Y as

$$Y = \underline{x}'\underline{\beta}^* + Z^* \, ,$$

where $Z^* = \beta_0^* + \sigma Z$, we see this model is an accelerated failure time model. Here $Z^* \sim$ extreme value with location β_0^* and scale σ. The linear predictor given on page 99 is

$$\tilde{\mu} = -\log(\tilde{\lambda}) = \beta_0^* + \underline{x}'\underline{\beta}^* \tag{4.21}$$

with $\beta_0^* = -\log(\lambda)$ and $\underline{\beta}^* = -\sigma\underline{\beta}$, where the vector $\underline{\beta}$ denotes the coefficients in the Weibull hazard on page 98 and, $\sigma = 1/\alpha$, where α denotes the Weibull shape parameter. Let $\widehat{\beta}_0^*$, $\widehat{\underline{\beta}}^{*\prime}$, and $\hat{\sigma}$ denote the MLE's of the parameters. Recall that the theory tells us MLE's are approximately normally distributed when the sample size n is large. To test $H_0 : \beta_j^* = \beta_j^{*0}$, $j = 1, \ldots, m$, use

$$\frac{\widehat{\beta}_j^* - \beta_j^{*0}}{\text{s.e.}(\widehat{\beta}_j^*)} \overset{a}{\sim} N(0,1) \quad \text{under } H_0.$$

An approximate $(1 - \alpha) \times 100\%$ confidence interval for β_j^* is given by

$$\widehat{\beta}_j^* \pm z_{\frac{\alpha}{2}} \text{ s.e.}(\widehat{\beta}_j^*),$$

where $z_{\frac{\alpha}{2}}$ is taken from the $N(0,1)$ table. Inferences concerning the intercept β_0^* follow analogously.

Notes:

1. It is common practice to construct $(1 - \alpha) \times 100\%$ confidence intervals for the coefficients in the Weibull model by multiplying both endpoints by

$-\widehat{\sigma}^{-1}$ and reversing their order. However, we suggest constructing confidence intervals using the bivariate delta method stated in Chapter 3.6 to obtain a more appropriate standard error for $\widehat{\beta}_j$. The reason is that the bivariate delta method takes into account the variability due to $\widehat{\sigma}$ as well as $\widehat{\beta}_j^*$. The common approach does not, and hence, could seriously underestimate the standard error. The explicit expression for the variance of $\widehat{\beta}_1$ is as follows:

$$\widehat{\text{var}}(\widehat{\beta}_1) = \frac{1}{\widehat{\sigma}^2}\left(\text{var}(\widehat{\beta}_1^*) + \widehat{\beta}_1^{*2}\text{var}(\log(\widehat{\sigma})) - 2\widehat{\beta}_1^*\text{cov}(\widehat{\beta}_1^*, \log(\widehat{\sigma}))\right). \quad (4.22)$$

WHY! We use this expression to compute a 95% confidence interval for β_1 at the end of this chapter.

2. It is common practice to compute a $(1-\alpha) \times 100\%$ confidence interval for the true parameter value of λ by multiplying LCL and UCL for the intercept β_0^* by -1, then taking the exp($\cdot$) of both endpoints, and then, reversing their order. This may end up with too wide a confidence interval as we show at the end of this chapter. Again we recommend the delta method to obtain the variance estimate of $\widehat{\lambda}$. By applying the delta method to $\widehat{\lambda} = \exp(-\widehat{\beta}_0^*)$, we obtain $\widehat{\text{var}}(\widehat{\lambda}) = \exp(-2\widehat{\beta}_0^*)\text{var}(\widehat{\beta}_0^*)$. WHY!

At the point $\underline{x} = \underline{x}_0$, the MLE of the $(p \times 100)$th percentile of the distribution of $Y = \log(T)$ is

$$\widehat{Y}_p = \widehat{\beta}_0^* + \underline{x}_0'\widehat{\underline{\beta}}^* + \widehat{\sigma}z_p = (1, \underline{x}_0', z_p)\begin{pmatrix} \widehat{\beta}_0^* \\ \widehat{\underline{\beta}}^* \\ \widehat{\sigma} \end{pmatrix},$$

where z_p is the $(p \times 100)$th percentile of the error distribution, which, in this case, is standard extreme value. The estimated variance of $\widehat{Y}_p$ is

$$\text{var}(\widehat{Y}_p) = (1, \underline{x}_0', z_p)\widehat{\Sigma}\begin{pmatrix} 1 \\ \underline{x}_0 \\ z_p \end{pmatrix}, \quad (4.23)$$

where $\widehat{\Sigma}$ is the estimated variance-covariance matrix of $\widehat{\beta}_0^*$, $\widehat{\beta}_1^*$, and $\widehat{\sigma}$. WHY! Then an approximate $(1 - \alpha) \times 100\%$ confidence interval for the $(p \times 100)$th percentile of the log-failure time distribution is given by

$$\widehat{Y}_p \pm z_{\frac{\alpha}{2}} \text{ s.e.}(\widehat{Y}_p),$$

where $z_{\frac{\alpha}{2}}$ is taken from the $N(0, 1)$ table. These are referred to as the **uquantile** type in the S function `predict`. The MLE of t_p is $\exp(\widehat{Y}_p)$. To obtain confidence limits for t_p, take the exponential of the endpoints of the above confidence interval.

The function `predict`, a companion function to `survReg`, conveniently reports both the quantiles in time and the uquantiles in log(time) along with their respective s.e.'s. We often find the confidence intervals based on uquantiles are shorter than those based on quantiles. See, for example, the results at the end of this section.

In S, we fit the model

$$Y = \log(\text{time}) = \beta_0^* + \beta_1^* x + \sigma Z,$$

where $Z \sim$ standard extreme value distribution. The $(p \times 100)$th percentile of the standard extreme (minimum) value distribution, Table 3.1, is

$$z_p = \log\left(-\log(1-p)\right).$$

The function survReg outputs the estimated variance-covariance matrix $\widehat{V}$ for the MLE's $\widehat{\beta_0^*}$, $\widehat{\beta_1^*}$, and $\widehat{\tau} = \log \widehat{\sigma}$. However, internally it computes $\widehat{\Sigma}$ to estimate the var$(\widehat{Y}_p)$.

The following is an S program along with modified output. The function survReg is used to fit a Weibull regression model. Then the 15th and 85th percentiles as well as the median failure time are estimated with corresponding standard errors. We also predict the failure time in hours at $x_0 = 2.480159$, which corresponds to the design temperature of $130^0 C$. Four plots of the estimated hazard and survivor functions are displayed in Figure 4.2. Three Q-Q plots are displayed in Figure 4.3, where intercept is $\widehat{\beta_0^*} + \widehat{\beta_1^*} x$ and slope is $\widehat{\sigma}$. Since there are three distinct values of x, we have three parallel lines. Lastly, the results are summarized.

```
> attach(motorette)
> weib.fit <- survReg(Surv(time,status)~x,dist="weibull")
> summary(weib.fit)
                Value Std. Error    z          p
(Intercept) -11.89      1.966   -6.05 1.45e-009
          x   9.04      0.906    9.98 1.94e-023
  Log(scale)  -1.02      0.220   -4.63 3.72e-006


> weib.fit$var # The estimated covariance matrix of the
             # coefficients and log(sigmahat).
             (Intercept)           x  Log(scale)
(Intercept)  3.86321759 -1.77877653  0.09543695
          x -1.77877653  0.82082391 -0.04119436
  Log(scale) 0.09543695 -0.04119436  0.04842333


> predict(weib.fit,newdata=list(x),se.fit=T,type="uquantile",
    p=c(0.15,0.5,0.85)) # newdata is required whenever
    # uquantile is used as a type whereas quantile uses the
    # regression variables as default. This returns the
    # estimated quantiles in log(t) along with standard
    # error as an option.
```

```
# Estimated quantiles in log(hours) and standard errors in
# parentheses. The output is edited because of redundancy.

x=2.256318    7.845713     8.369733      8.733489
             (0.1806513)  (0.12339772)  (0.1370423)
x=2.158895    6.965171     7.489190      7.852947
             (0.1445048)  (0.08763456)  (0.1189669)
x=2.027575    5.778259     6.302279      6.666035
             (0.1723232)  (0.14887233)  (0.1804767)


> predict(weib.fit,newdata=data.frame(x=2.480159),se.fit=T,
    type="uquantile",p=c(0.15,0.5,0.85)) # Estimated
    # quantiles in log(hours) at the new x value =
    # 2.480159; i.e., the design temperature of 130
    # degrees Celsius.

x=2.480159    9.868867     10.392887     10.756643
             (0.3444804)  (0.3026464)   (0.2973887)


> sigmahat <- weib.fit$scale
> alphahat <- 1/sigmahat # estimate of shape
> coef <- weib.fit$coef
> lambdatildehat <- exp(- coef[1] - coef[2]*2.480159)
    # estimate of scale
> pweibull(25000,alphahat,1/lambdatildehat) # Computes the
    # estimated probability that a motorette failure time
    # is less than or equal to 25,000 hours. pweibull is
    # the Weibull distribution function in S.

  [1] 0.2783054 # estimated probability

> Shatq <- 1 - 0.2783054 # survival probability at 25,000
    # hours. About 72% of motorettes are still working
    # after 25,000 hours at x=2.480159; i.e., the design
    # temperature of 130 degrees Celsius.

> xl <- levels(factor(x)) # Creates levels out of the
                          # distinct x-values.
> ts.1 <- Surv(time[as.factor(x)==xl[1]],
            status[as.factor(x)==xl[1]]) # The first
                                # group of data
> ts.2 <- Surv(time[as.factor(x)==xl[2]],
            status[as.factor(x)==xl[2]]) # The second
> ts.3 <- Surv(time[as.factor(x)==xl[3]],
            status[as.factor(x)==xl[3]]) # The third
```

```
> par(mfrow=c(2,2)) # divides a screen into 2 by 2 pieces.
> Svobj <- list(ts.1,ts.2,ts.3) # Surv object
> qq.weibreg(Svobj,weib.fit) # The first argument takes
        # a Surv object and the second a survReg object.
        # Produces a Weibull Q-Q plot.
> qq.loglogisreg(Svobj,loglogistic.fit) # log-logistic
                                        # Q-Q plot
> qq.lognormreg(Svobj,lognormal.fit) # log-normal Q-Q plot
> detach()
```

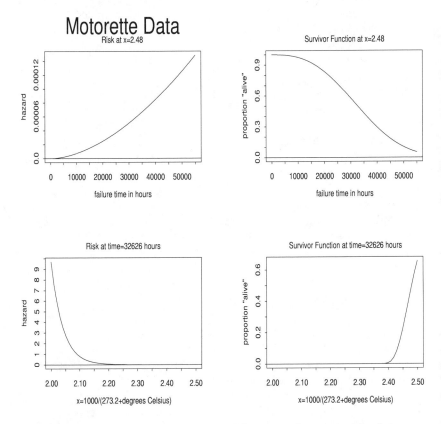

Motorette Data

Figure 4.2 *Weibull hazard and survival functions fit to motorette data.*

Results:

- From summary(weib.fit), we learn that $\widehat{\sigma} = \exp(-1.02) = .3605949$, and $\widehat{\mu} = -\log(\widehat{\lambda}) = \widehat{\beta_0^*} + \widehat{\beta_1^*}x = -11.89 + 9.04x$.

  Thus, we obtain $\widehat{\alpha} = \frac{1}{.3605949} = 2.773195$ and $\widehat{\lambda} = \exp(11.89 - 9.04 \times$

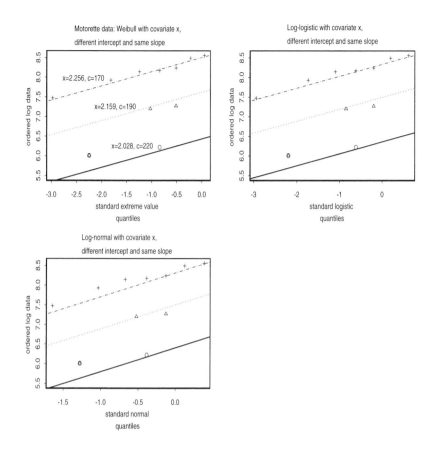

Figure 4.3 *Weibull, log-logistic, and log-normal Q-Q plots of the motorette data. Lines constructed with MLE's.*

2.480159$) = 0.0000267056$ at $x = 2.480159$. Note also that both the intercept and covariate x are highly significant with p-values 1.45×10^{-9} and 1.94×10^{-23}, respectively.

- It follows from Chapter 4.2 that the estimated hazard function is

$$\widehat{h}(t|x) = \frac{1}{\widehat{\sigma}} \cdot t^{\frac{1}{\widehat{\sigma}}-1} \cdot (\exp(-\widehat{\mu}))^{\frac{1}{\widehat{\sigma}}}$$

and the estimated survivor function is

$$\widehat{S}(t|x) = \exp\left\{ -\left(\exp(-\widehat{\mu})t \right)^{\frac{1}{\widehat{\sigma}}} \right\}.$$

- The point estimate of β_1, $\widehat{\beta}_1$, is $-\widehat{\sigma}^{-1}\widehat{\beta}_1^*$. A 95% C.I. for β_1 based on the delta method is given by $[-37.84342, -12.29594]$. Whereas the one based on the common approach is given by

$$[-\hat{\sigma}^{-1}(10.82), -\hat{\sigma}^{-1}(7.26)] = [-29.92, -20.09],$$

where $\hat{\sigma} = .3605949$ and the 95% C.I. for β_1^* is $[7.26, 10.81] = [9.04 - 1.96 \times .906, 9.04 + 1.96 \times .906]$. It is clear that the latter interval is much shorter than the former as it ignores the variability of $\hat{\sigma}$.

- A 95% C.I. for λ based on the delta method is given by $[-416023.7, 707626.3]$. We see this includes negative values, which is not appropriate because λ is restricted to be positive. Therefore, we report the truncated interval $[0, 707626.3]$. The one based on the common approach is given by

$$[\exp(8.04), \exp(15.74)] = [3102.61, \ 6851649.6],$$

where the 95% C.I. for β_0^* is $[-11.89 - 1.96 \times 1.966, -11.89 + 1.96 \times 1.966] = [-15.74, -8.04]$. Although the common approach ends up with an unreasonably wide confidence interval compared to the one based on the delta method, this approach always yields limits within the legal range of λ.

- At $x = 2.480159$, the design temperature of 130^0C, the estimated 15th, 50th, and 85th percentiles in log(hours) and hours, respectively based on uquantile and quantile, along with their corresponding 90% C.I.'s in hours are reported in the following table.

type	percentile	Estimate	Std.Err	90% LCL	90% UCL
uquantile	15	9.868867	0.3444804	10962.07	34048.36
	50	10.392887	0.3026464	19831.64	53677.02
	85	10.756643	0.2973887	28780.08	76561.33
quantile	15	19319.44	6655.168	9937.174	37560.17
	50	32626.7612	9874.361	19668.762	54121.65
	85	46940.83	13959.673	28636.931	76944.21

The 90% C.I.'s based on uquantile, $\exp(\text{estimate} \pm 1.645 \times \text{std.err})$, are shorter than those based on quantile at each x value. However, we also suspect there is a minor bug in predict in that there appears to be a discrepancy between the standard error estimate for the 15th percentile resulting from uquantile and ours based on the delta method which follows. The other two standard error estimates are arbitrarily close to ours. Our standard error estimates are .3174246, .2982668, and .3011561 for the 15th, 50th, and 85th percentiles, respectively. Applying the trivariate delta method, we obtain the following expression:

$$\widehat{\text{var}}(\hat{y}_p) = \text{var}(\widehat{\beta_0^*}) + \text{var}(\widehat{\beta_1^*})x_0^2 + z_p^2\hat{\sigma}^2\text{var}(\log\hat{\sigma}) \qquad (4.24)$$
$$+ \ 2x_0\text{cov}(\widehat{\beta_0^*}, \widehat{\beta_1^*}) + 2z_p\hat{\sigma}\text{cov}(\widehat{\beta_0^*}, \log\hat{\sigma}) + 2x_0z_p\hat{\sigma}\text{cov}(\widehat{\beta_1^*}, \log\hat{\sigma}).$$

WHY!

- At the design temperature 130^0C, by 25,000 hours about 28% of the motorettes have failed. That is, after 25,000 hours, about 72% are still working.

- As $\hat{\alpha} = \frac{1}{\hat{\sigma}} = \frac{1}{.3605949} = 2.773195$, then for fixed x the hazard function increases as time increases. The upper two graphs in Figure 4.2 display estimated hazard and survivor functions. The covariate x is fixed at 2.480159 which corresponds to the design temperature 130^0C.

- The estimated coefficient $\hat{\beta}_1 = -\frac{1}{\hat{\sigma}}\hat{\beta}_1^* = -\frac{1}{.3605949}(9.04) = -25.06968 < 0$. Thus, for time fixed, as x increases, the hazard decreases and survival increases. The lower two graphs in Figure 4.2 display these estimated functions when time is fixed at 32,626 hours.

- For $x_1 < x_2$,
$$\frac{h(t|x_2)}{h(t|x_1)} = \exp((x_2 - x_1)(-25.06968)).$$
For example, for $x = 2.1$ and 2.2,
$$\frac{h(t|2.2)}{h(t|2.1)} = \exp(.1(-25.06968)) = .08151502.$$
Thus, for .1 unit increase in x, the hazard becomes about 8.2% of the hazard before the increase. In terms of Celsius temperature, for 21.645 degree decrease from 202.9905^0C to 181.3455^0C, the hazard becomes about 8.2% of the hazard before the decrease.

- The Q-Q plots in Figure 4.3 show that the Weibull fit looks slightly better than the log-logistic fit at the temperature 170^0C, but overall they are the same. On the other hand, the Weibull fit looks noticeably better than the log-normal fit at the temperature 170^0C and is about the same at the other two temperatures. This result coincides with our finding from AIC in Table 4.2; that is, among these three accelerated failure time models, the Weibull best describes the motorette data.

4.7 Exercises

A. *Applications*

4.1 We work with the diabetes data set again. Refer to Exercise 2.3. Consider the Weibull regression model
$$Y = \log(\texttt{lzeit}) = \beta_0^* + \beta_1^* x^{(1)} + \beta_2^* x^{(2)} + \beta_3^* x^{(3)} + \sigma Z,$$
where $Z \sim$ standard extreme value and
$$x^{(1)} = \begin{cases} 0 & \text{man} \\ 1 & \text{woman} \end{cases}$$

$$x^{(2)} = \begin{cases} 0 & \text{nondiabetic} \\ 1 & \text{diabetic} \end{cases}$$

$$x^{(3)} = \text{age in years}$$

(a) Estimate σ and the coefficients β_j^*. Which covariates are significant?
   Tips:

   ```
   > fit.a <- survReg(Surv(lzeit,tod) ~ sex+diab+alter,
                       dist="weibull",data=diabetes)
   > summary(fit.a)
   ```

(b) We now add two additional covariates which models in the possible dependence of diabetic or not with age. That is, we replace $x^{(3)}$ with the following interaction variables:

$$x^{(4)} = \begin{cases} \text{age, if diabetic} \\ 0 \qquad \text{otherwise} \end{cases}$$

$$x^{(5)} = \begin{cases} \text{age, if nondiabetic} \\ 0 \qquad \text{otherwise} \end{cases}$$

   Describe the results of the analysis with the four covariates now. Which covariates are significant?
   Tips:

   ```
   > diabetes$x4 <- diabetes$x5 <- diabetes$alter
   > diabetes$x4[diabetes$diab==0] <- 0
   > diabetes$x5[diabetes$diab==1] <- 0
   > fit.b <- survReg(Surv(lzeit,tod) ~ sex+diab+x4+x5,
                       data=diabetes)
   > summary(fit.b)
   ```

(c) Simplify the model fit in part (b) as much as possible. Draw conclusions (as much as possible as you are not diabetes specialists, etc.).

   For the remaining parts, use the fitted additive model `fit.a` (part (a)) with just `sex`, `diab`, and `alter` in the model.

(d) Report the estimated hazard function for those who are men and nondiabetic.

   Tip:

   See the summary on page 105.

(e) Report the estimated hazard ratio comparing diabetic men to nondiabetic men all of whom have the same age. Interpret this ratio.

(f) A 50-year-old nondiabetic man is operated on today. What is the estimated probability that he is still alive in ten years?

(g) With the help of the `predict` function, calculate the survival duration (in days or years) after the operation within which half (50%) of the 50-year-old diabetic men have died and then, similarly, for nondiabetic men. Report both point and interval estimates.

Tips:

Use the preferred C.I. approach (Table 3.2 on Chapter 3); that is, `type= "uquantile"`. Use the help routine in S or R to look up `predict`. Be sure to study the example given there. See the S code to compute the medhat for Model 2 on page 85.

4.2 In order to better understand the age dependence of survival **"lzeit"**, plot now the survival times against "$x^{(5)}$" and then against "$x^{(4)}$". Comment. Is there something here that helps explain what it is you are observing?

Tips:

To investigate the age structure inherent in the raw data set, split the data set in to two sets: one with data corresponding to diabetics, the other with nondiabetics.

```
> nondiab <- diabetes[diab==0, ]
> diab <- diabetes[diab==1, ]
> par(mfrow=c(1,2))
> plot(nondiab$alter,nondiab$lzeit,type="none",
          ylim=range(diabetes$lzeit),ylab="Survival duration",
          xlab="Alter")
> text(jitter(nondiab$alter),jitter(nondiab$lzeit),
          labels=nondiab$tod)
> lines(smooth.spline(nondiab$alter,nondiab$lzeit))
> plot(diab$alter,diab$lzeit,type="none",
          ylim=range(diabetes$lzeit),ylab="Survival duration",
          xlab="Alter")
> text(jitter(diab$alter),jitter(diab$lzeit),labels=diab$tod)
> lines(smooth.spline(diab$alter,diab$lzeit))
```

4.3 This problem parallels Exercise 4.1. We now consider fitting the log-logistic regression model.

(a) Repeat part (a), (b), and (c).

For the remaining parts, use the fitted additive model with just `sex`, `diab`, and `alter` in the model.

(b) Report the estimated odds-ratio OR for those who are men and nondiabetic. at arbitrary two different ages.

(c) Report the estimated odds of survival beyond time t.

(d) Report the estimated odds-ratio OR comparing diabetic men to nondiabetic men all of whom have the same age. Interpret this ratio.

(e) Report the estimated times-ratio TR comparing diabetic men to nondiabetic men all of whom have the same age. Interpret this ratio.

(f) A 50-year-old nondiabetic man is operated on today. What is the estimated probability that he is still alive in ten years?

(g) With the help of `predict` function, calculate the survival duration (in days or years) after the operation within which half (50%) of the 50-year-old diabetic men have died and then, similarly, for nondiabetic men. Report both point and interval estimates.

Tips:

Use the preferred C.I. approach (Table 3.2 on Chapter 3); that is, `type= "uquantile"`. See the summary on page 105.

B. *Theory and WHY!*

4.4 Verify expression (4.10).

4.5 Prove expression (4.14).

4.6 Verify expression (4.15).

4.7 Derive expression (4.18).

4.8 Derive expression (4.22).

4.9 Answer the WHY! at the second item under **Notes:** on page 111.

4.10 Derive expression (4.23).

4.11 Verify expression (4.24).

## CHAPTER 5

The Cox Proportional Hazards Model

In this chapter we discuss some features of a prognostic factor analysis based on the Cox proportional hazards (PH) model. We present an **analysis of the CNS lymphoma data** introduced in Example 2 in Chapter 1.1. The primary endpoint of interest here is survival time (in years) from first blood brain barrier disruption (BBBD) to death (B3TODEATH). Some questions of interest are:

1. Is there a difference in survival between the two groups (prior radiation, no radiation prior to first BBBD)?

2. Do any subsets of available covariates help explain this survival time? For example, does age at time of first treatment and/or gender increase or decrease the hazard of death; hence, decrease or increase the probability of survival; and hence, decrease or increase mean or median survival time?

3. Is there a dependence of the difference in survival between the groups on any subset of the available covariates?

Objectives of this chapter:

After studying Chapter 5, the student should:

1. Know and understand the definition of a Cox PH model including the assumptions.

2. Know how to use the S function `coxph` to fit data to a Cox PH model.

3. Know how to use the S function `stepAIC` along with `coxph` to identify an appropriate model.

4. Know how to use the **stratified Cox PH model**.

5. Know how to interpret the estimated β coefficients with respect to hazard and other features of the distribution.

6. Understand how to interpret the estimated hazards ratio HR. That is, understand its usefulness as a measure of effect that describes the relationship between the predictor variable(s) and time to failure. Further, the HR can be used to examine the relative likelihood of survival.

We first plot the two Kaplan-Meier (K-M) survivor curves using S. This plot displays a difference in survival between the two groups. The higher K-M curve for the no prior radiation group suggests that this group has a higher chance of long term survival. The following S output confirms this. The S function survdiff yields a **log-rank** test statistic value of 9.5 which confirms this difference with an approximate p-value of .002. Further note the estimated mean and median given in the output from the S function survfit. Much of the output has been deleted where not needed for discussion. The CNS data is stored in a data frame named cns2.

```
> cns2.fit0 <- survfit(Surv(B3TODEATH,STATUS)~GROUP,data=cns2,
      type="kaplan-meier")
> plot(cns2.fit0,lwd=3,col=1,type="l",lty=c(1,3),cex=2,
      lab=c(10,10,7),xlab="Survival Time in Years from
      First BBBD",ylab="Percent Surviving",yscale=100)
> text(6,1,"Primary CNS Lymphoma Patients",lwd=3)
> legend(3,0.8,type="l",lty=c(1,3,0),c("no radiation prior
      to BBBD (n=39)","radiation prior to BBBD (n=19)",
      "+ = patient is censored"),col=1)
```

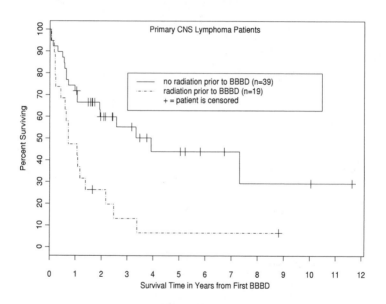

Figure 5.1 *Kaplan-Meier survivor curves.*

```
> survdiff(Surv(B3TODEATH,STATUS)~GROUP,data=cns2)
```

	N	Observed	Expected	(O-E)^2/E	(O-E)^2/V
GROUP=0	39	19	26.91	2.32	9.52
GROUP=1	19	17	9.09	6.87	9.52

Chisq= 9.5 on 1 degrees of freedom, p= 0.00203

```
> cns2.fit0
```

	n	events	mean	se(mean)	median	0.95LCL	0.95UCL
GROUP=0	39	19	5.33	0.973	3.917	1.917	NA
GROUP=1	19	17	1.57	0.513	0.729	0.604	2.48

Since the two survival curves are significantly different, we assess the factors that may play a role in survival and in this difference in survival duration. Recall that the hazard (risk) function, $h(t)\Delta t$, is approximately the conditional probability of failure in the (small) interval from t to $t+\Delta t$ given survival until time t. Here t is the length of time a patient lives from the point of his/her first BBBD. **Assuming that the baseline hazard function is the same for all patients in the study, a Cox PH model** seems appropriate. That is, we model the hazard rate as a function of the covariates $\underline{x}$. Recall from Chapter 4.3 that the **hazard function** has the form

$$h(t|\underline{x}) = h_0(t) \cdot \exp(\underline{x}'\underline{\beta}) = h_0(t) \cdot \exp\left(\beta_1 x^{(1)} + \beta_2 x^{(2)} + \cdots + \beta_m x^{(m)}\right)$$

$$= h_0(t) \cdot \exp\left(\beta_1 x^{(1)}\right) \times \exp\left(\beta_2 x^{(2)}\right) \cdots \times \exp\left(\beta_m x^{(m)}\right),$$

where $h_0(t)$ is an unspecified baseline hazard function free of the covariates $\underline{x}$. The covariates act multiplicatively on the hazard. At two different points $\underline{x}_1$ and $\underline{x}_2$, the proportion

$$\frac{h(t|\underline{x}_1)}{h(t|\underline{x}_2)} = \frac{\exp(\underline{x}_1'\underline{\beta})}{\exp(\underline{x}_2'\underline{\beta})}$$

$$= \frac{\exp\left(\beta_1 x_1^{(1)}\right) \times \exp\left(\beta_2 x_1^{(2)}\right) \times \cdots \times \exp\left(\beta_m x_1^{(m)}\right)}{\exp\left(\beta_1 x_2^{(1)}\right) \times \exp\left(\beta_2 x_2^{(2)}\right) \times \cdots \times \exp\left(\beta_m x_2^{(m)}\right)}$$

is constant with respect to time t. As we are interested in estimating the coefficients β, the baseline hazard is really a nuisance parameter. Through the **partial likelihood** (Cox, 1975) we obtain estimates of the coefficients β without regard to the baseline hazard $h_0(t)$. Note that in the parametric regression setting of Chapter 4, we specify the form of this function since we must specify a distribution for the target variable T. Chapter 8 presents a method for analyzing T, or $Y = \log(T)$, directly without assuming a specific distributional form. Remember that the hazard function completely specifies the distribution of T; but the power of the PH model is that it provides a fairly wide family of distributions by allowing the baseline hazard $h_0(t)$ to be arbitrary. The S function `coxph` implements Cox's partial likelihood function. In Chapter 6.3 we offer a heuristic derivation of Cox's partial likelihood.

5.1 AIC procedure for variable selection

Akaike's information criterion (AIC) for the Cox PH model:

We revisit AIC in the context of the Cox PH regression model. Comparisons between a number of possible models can be made on the basis of the statistic

$$AIC = -2 \times \log(\text{maximum likelihood}) + 2 \times b, \qquad (5.1)$$

where b is the number of β coefficients in each model under consideration. The maximum likelihood is replaced by the maximum partial likelihood. The smaller the AIC value the better the model is.

We apply an automated model selection procedure via an S function stepAIC included in MASS, a collection of functions and data sets from *Modern Applied Statistics with S* by Venables and Ripley (2002). Otherwise, it would be too tedious because of many steps involved.

The stepAIC function requires an object representing a model of an appropriate class. This is used as the initial model in the stepwise search. Useful optional arguments include scope and direction. The scope defines the range of models examined in the stepwise search. The direction can be one of "both," "backward," or "forward," with a default of "both." If the scope argument is missing, the default for direction is "backward." We illustrate how to use stepAIC together with LRT to select a best model. We consider an example fitting CNS data to Cox PH model.

Example:

For ease of reading, we reprint variable code.

1. PT.NUMBER: patient number
2. GROUP: 0=no prior radiation with respect to 1st blood brain barrier disruption (BBBD) procedure to deliver chemotherapy ; 1=prior radiation
3. SEX: 0=male ; 1=female
4. AGE: at time of 1st BBBD, recorded in years
5. STATUS: 0=alive ; 1=dead
6. DXTOB3: time from diagnosis to 1st BBBD in years
7. DXTODeath: time from diagnosis to death in years
8. B3TODeath: time from 1st BBBD to death in years
9. KPS.PRE.: Karnofsky performance score before 1st BBBD, numerical value 0 − 100
10. LESSING: Lesions: single=0 ; multiple=1
11. LESDEEP: Lesions: superficial=0 ; deep=1
12. LESSUP: Lesions: supra=0 ; infra=1 ; both=2
13. PROC: Procedure: subtotal resection=1 ; biopsy=2 ; other=3
14. RAD4000: Radiation > 4000: no=0 ; yes=1
15. CHEMOPRIOR: no=0 ; yes=1
16. RESPONSE: Tumor response to chemotherapy - complete=1; partial=2; blanks represent missing data

In Chapter 1.2 we established the relationship that the smaller the risk, the larger the probability of survival, and hence the greater mean survival.

The estimates from fitting a Cox PH model are interpreted as follows:

• A positive coefficient increases the risk and thus decreases the expected (average) survival time.
• A negative coefficient decreases the risk and thus increases the expected survival time.
• The ratio of the estimated risk functions for the two groups can be used to examine the likelihood of Group 0's (no prior radiation) survival time being longer than Group 1's (with prior radiation).

The two covariates LESSUP and PROC are categorical. Each has three levels. The S function `factor` creates indicator variables. Also, the variable AGE60 is defined as AGE60 = 1 if AGE $\leq$ 60 and = 0 otherwise. We implement the S functions `stepAIC` and `coxph` to select appropriate variables according to the AIC criterion based on the proportional hazards model.

Let us consider the two-way interaction model, which can be easily incorporated in the `stepAIC`. Three-way or four-way interaction models can be considered but the interpretation of an interaction effect, if any, is not easy. The initial model contains all 11 variables without interactions. The scope is up to two-way interaction models. These are listed in the S code under Step I that follows. The direction is "both." The AIC for each step is reported in Table 5.1. The first AIC value is based on the initial model of 11 variables without interactions. "+" means that term was added at that step and "-" means that term was removed at that step. The final model retains the following variables: KPS.PRE., GROUP, SEX, AGE60, LESSING, CHEMO-PRIOR, SEX:AGE60, AGE60:LESSING, and GROUP:AGE60.

Step I: stepAIC to select the best model according to AIC statistic

```
> library(MASS) # Call in a collection of library functions
                # provided by Venables and Ripley
> attach(cns2)
> cns2.coxint<-coxph(Surv(B3TODEATH,STATUS)~KPS.PRE.+GROUP+SEX+
  AGE60+LESSING+LESDEEP+factor(LESSUP)+factor(PROC)+CHEMOPRIOR)
          # Initial model              2x2 interactions
> cns2.coxint1 <- stepAIC(cns2.coxint,~.^2)
                # Up to two-way interaction
> cns2.coxint1$anova # Shows stepwise model path with the
                # initial and final models
```

Table 5.1: *Stepwise model path for
two-way interaction model on the CNS
lymphoma data*

Step	Df	AIC
		246.0864
+ SEX:AGE60	1	239.3337
- factor(PROC)	2	236.7472
- LESDEEP	1	234.7764
- factor(LESSUP)	2	233.1464
+ AGE60:LESSING	1	232.8460
+ GROUP:AGE60	1	232.6511

Step II: LRT to further reduce

The following output shows p-values corresponding to variables selected by
stepAIC. AGE60 has a large p-value, .560, while its interaction terms with
SEX and LESSING have small p-values, .0019 and .0590, respectively.

```
> cns2.coxint1 # Check which variable has a
               # moderately large p-value

                  coef exp(coef) se(coef)      z       p
      KPS.PRE. -0.0471    0.9540    0.014 -3.362 0.00077
         GROUP  2.0139    7.4924    0.707  2.850 0.00440
           SEX -3.3088    0.0366    0.886 -3.735 0.00019
         AGE60 -0.4037    0.6679    0.686 -0.588 0.56000
       LESSING  1.6470    5.1916    0.670  2.456 0.01400
    CHEMOPRIOR  1.0101    2.7460    0.539  1.876 0.06100
     SEX:AGE60  2.8667   17.5789    0.921  3.113 0.00190
 AGE60:LESSING -1.5860    0.2048    0.838 -1.891 0.05900
   GROUP:AGE60 -1.2575    0.2844    0.838 -1.500 0.13000
```

In statistical modelling, an important principle is that an interaction term
should only be included in a model when the corresponding main effects are
also present. We now see if we can eliminate the variable AGE60 and its
interaction terms with other variables. We use the LRT. Here the LRT is
constructed on the partial likelihood function rather than the full likelihood
function. Nonetheless the large sample distribution theory holds. The LRT
test shows strong evidence against the reduced model and so we retain the
model selected by stepAIC.

```
> cns2.coxint2 <- coxph(Surv(B3TODEATH,STATUS)~KPS.PRE.+GROUP
        +SEX+LESSING+CHEMOPRIOR) # Without AGE60 and its
                                 # interaction terms
> -2*cns2.coxint2$loglik[2] + 2*cns2.coxint1$loglik[2]
[1] 13.42442
```

```
> 1 - pchisq(13.42442,4)
[1] 0.009377846 # Retain the model selected by stepAIC
```

Now we begin the process of one variable at a time reduction. This can be based on either the p-value method or the LRT. Asymptotically they are equivalent. Since the variable GROUP:AGE60 has a moderately large p-value, .130, we delete it. The following LRT test shows no evidence against the reduced model (p-value = .138) and so we adopt the reduced model.

```
> cns2.coxint3 <- coxph(Surv(B3TODEATH,STATUS)~KPS.PRE.+GROUP
    +SEX+AGE60+LESSING+CHEMOPRIOR+SEX:AGE60+AGE60:LESSING)
      # Without GROUP:AGE60
> -2*cns2.coxint3$loglik[2] + 2*cns2.coxint1$loglik[2]
[1] 2.194949
> 1 - pchisq(2.194949,1)
[1] 0.1384638 # Selects the reduced model

> cns2.coxint3 # Check which variable has a
              # moderately large p-value
```

	coef	exp(coef)	se(coef)	z	p
KPS.PRE.	-0.0436	0.9573	0.0134	-3.25	0.0011
GROUP	1.1276	3.0884	0.4351	2.59	0.0096
SEX	-2.7520	0.0638	0.7613	-3.61	0.0003
AGE60	-0.9209	0.3982	0.5991	-1.54	0.1200
LESSING	1.3609	3.8998	0.6333	2.15	0.0320
CHEMOPRIOR	0.8670	2.3797	0.5260	1.65	0.0990
SEX:AGE60	2.4562	11.6607	0.8788	2.79	0.0052
AGE60:LESSING	-1.2310	0.2920	0.8059	-1.53	0.1300

From this point on we use the p-value method to eliminate one term at a time. As AGE60:LESSING has a moderately large p-value, .130, we remove it.

```
> cns2.coxint4 # Check which variable has a
              # moderately large p-value
```

	coef	exp(coef)	se(coef)	z	p
KPS.PRE.	-0.0371	0.9636	0.0124	-3.00	0.00270
GROUP	1.1524	3.1658	0.4331	2.66	0.00780
SEX	-2.5965	0.0745	0.7648	-3.40	0.00069
AGE60	-1.3799	0.2516	0.5129	-2.69	0.00710
LESSING	0.5709	1.7699	0.4037	1.41	0.16000
CHEMOPRIOR	0.8555	2.3526	0.5179	1.65	0.09900
SEX:AGE60	2.3480	10.4643	0.8765	2.68	0.00740

We eliminate the term LESSING as it has a moderately large p-value, .160.

```
> cns2.coxint5 # Check which variable has a
              # moderately large p-value
```

	coef	exp(coef)	se(coef)	z	p
KPS.PRE.	-0.0402	0.9606	0.0121	-3.31	0.00093
GROUP	0.9695	2.6366	0.4091	2.37	0.01800
SEX	-2.4742	0.0842	0.7676	-3.22	0.00130
AGE60	-1.1109	0.3293	0.4729	-2.35	0.01900
CHEMOPRIOR	0.7953	2.2152	0.5105	1.56	0.12000
SEX:AGE60	2.1844	8.8856	0.8713	2.51	0.01200

We eliminate the variable CHEMOPRIOR as it has a moderately large p-value, .120. Since all the p-values in the reduced model fit below are small enough at the .05 level, we finally stop here and retain these five variables: KPS.PRE., GROUP, SEX, AGE60, and SEX:AGE60.

```
> cns2.coxint6 # Check which variable has a
              # moderately large p-value
```

	coef	exp(coef)	se(coef)	z	p
KPS.PRE.	-0.0307	0.970	0.0102	-2.99	0.0028
GROUP	1.1592	3.187	0.3794	3.06	0.0022
SEX	-2.1113	0.121	0.7011	-3.01	0.0026
AGE60	-1.0538	0.349	0.4572	-2.30	0.0210
SEX:AGE60	2.1400	8.500	0.8540	2.51	0.0120

However, it is important to compare this model to the model chosen by stepAIC in Step I as we have not compared them. The p-value based on LRT is between .05 and .1 and so we select the reduced model with caution.

```
> -2*cns2.coxint6$loglik[2] + 2*cns2.coxint1$loglik[2]
[1] 8.843838
> 1 - pchisq(8.843838,4)
[1] 0.06512354 # Selects the reduced model
```

The following output is based on the model with KPS.PRE., GROUP, SEX, AGE60, and SEX:AGE60. It shows that the three tests – LRT, Wald, and efficient score test – indicate there is an overall significant relationship between this set of covariates and survival time. That is, they are explaining a significant portion of the variation.

```
> summary(cns2.coxint6)
```

```
Likelihood ratio test= 27.6  on 5 df,   p=0.0000431
Wald test            = 24.6  on 5 df,   p=0.000164
Score (logrank) test = 28.5  on 5 df,   p=0.0000296
```

This model is substantially different from that reported in Dahlborg *et al.* (1996). We go through model diagnostics in Chapter 6 to confirm that the model does fit the data.

Remarks:

1. The model selection procedure may well depend on the purpose of the study. In some studies there may be a few variables of special interest. In this case, we can still use Step I and Step II. In Step I we select the best set of variables according to the smallest AIC statistic. If this set includes all the variables of special interest, then in Step II we have only to see if we can further reduce the model. Otherwise, add to the selected model the unselected variables of special interest and go through Step II.

2. It is important to include interaction terms in model selection procedures unless researchers have compelling reasons why they do not need them. As the following illustrates, we could end up with a quite different model when only main effects models are considered.

   We reexamine the CNS Lymphoma data. The AIC for each model without interaction terms is reported in Table 5.2. The first AIC is based on the initial model including all the variables. The final model is selected by applying backward elimination procedure with the range from the full model with all the variables to the smallest reduced model with intercept only. It retains these four variables: KPS.PRE., GROUP, SEX, and CHEMO-PRIOR.

Step I: stepAIC to select the best model according to AIC statistic

```
> cns2.cox <- coxph(Surv(B3TODEATH,STATUS)~KPS.PRE.+GROUP+SEX
    +AGE60+LESSING+LESDEEP+factor(LESSUP)+factor(PROC)
    +CHEMOPRIOR) # Initial model with all variables
> cns2.cox1 <- stepAIC(cns2.cox,~.) # Backward elimination
            # procedure from full model to intercept only
> cns2.cox1$anova # Shows stepwise model paths with the
            # initial and final models
```

just Main effects

Table 5.2: *Stepwise model path for the main effects model*

Step	Df	AIC
		246.0864
- factor(PROC)	2	242.2766
- LESDEEP	1	240.2805
- AGE60	1	238.7327
- factor(LESSUP)	2	238.0755
- LESSING	1	236.5548

Step II: LRT to further reduce

The following output shows p-values corresponding to variables selected by stepAIC. The p-values corresponding to GROUP and CHEMOPRIOR are very close. This implies that their effects adjusted for the other variables are about the same.

```
> cns2.cox1 # Check which variable has a large p-value
```

	coef	exp(coef)	se(coef)	z	p
KPS.PRE.	-0.0432	0.958	0.0117	-3.71	0.00021
GROUP	0.5564	1.744	0.3882	1.43	0.15000
SEX	-1.0721	0.342	0.4551	-2.36	0.01800
CHEMOPRIOR	0.7259	2.067	0.4772	1.52	0.13000

We first eliminate GROUP. Since all the p-values in the reduced model are small enough at .05 level, we finally stop here and retain these three variables: KPS.PRE., SEX, and CHEMOPRIOR.

```
> cns2.cox2 # Check which variable has a
           # moderately large p-value
```

	coef	exp(coef)	se(coef)	z	p
KPS.PRE.	-0.0491	0.952	0.011	-4.46	8.2e-006
SEX	-1.2002	0.301	0.446	-2.69	7.1e-003
CHEMOPRIOR	1.0092	2.743	0.440	2.30	2.2e-002

Now let us see what happens if we eliminate CHEMOPRIOR first instead of GROUP. Since all the p-values in the reduced model are either smaller or about the same as .05 level, we stop here and retain these three variables: KPS.PRE., GROUP, and SEX.

```
> cns2.cox3 # Check which variable has large p-value
```

	coef	exp(coef)	se(coef)	z	p
KPS.PRE.	-0.0347	0.966	0.010	-3.45	0.00056
GROUP	0.7785	2.178	0.354	2.20	0.02800
SEX	-0.7968	0.451	0.410	-1.94	0.05200

```
> detach()
```

In summary, depending on the order of elimination, we retain either SEX, KPS.PRE., and CHEMOPRIOR, or KPS.PRE., GROUP, and SEX. These two models are rather different in that one includes CHEMOPRIOR where the other includes GROUP instead. More importantly, note that none of these sets include the variable AGE60, which is a very important prognostic factor in this study evidenced by its significant interaction effect with SEX on the response (cns2.coxint6). In addition, the significance of the GROUP effect based on the interaction model is more pronounced (p-value 0.0022 versus 0.028), which was the primary interest of the study. Therefore, we choose the interaction model cns2.coxint6 on page 128 to discuss.

Discussion

- KPS.PRE., GROUP, SEX, AGE60, and SEX:AGE60 appear to have a significant effect on survival duration. Here it is confirmed again that there is a significant difference between the two groups' (0=no prior radiation,1=prior radiation) survival curves.

- The estimated coefficient for KPS.PRE. is $-.0307$ with p-value 0.0028. Hence, fixing other covariates, patients with high KPS.PRE. scores have a decreased hazard, and, hence, have longer expected survival time than those with low KPS.PRE. scores.

- The estimated coefficient for GROUP is 1.1592 with p-value 0.0022. Hence, with other covariates fixed, patients with radiation prior to first BBBD have an increased hazard, and, hence, have shorter expected survival time than those in Group 0.

- Fixing other covariates, the hazard ratio between Group 1 and Group 0 is

$$\frac{\exp(1.1592)}{\exp(0)} = 3.187.$$

This means that, with other covariates fixed, patients with radiation prior to first BBBD are 3.187 times more likely than those without to have shorter survival.

- Fixing other covariates, if a patient in Group 1 has 10 units larger KPS.PRE. score than a patient in Group 0, the ratio of hazard functions is

$$\frac{\exp(1.1592)\exp(-0.0307 \times (k+10))}{\exp(0)\exp(-.0307 \times k)} = \frac{\exp(1.1592)\exp(-0.0307 \times 10)}{\exp(0)}$$
$$= 3.187 \times 0.7357 = 2.345,$$

where k is an arbitrary number. This means that fixing other covariates, a patient in Group 1 with 10 units larger KPS.PRE. score than a patient in Group 0 is 2.34 times more likely to have shorter survival. In summary, fixing other covariates, whether a patient gets radiation therapy prior to first BBBD is more important than how large his/her KPS.PRE. score is.

- There is significant interaction between AGE60 and SEX. The estimated coefficient for SEX:AGE60 is 2.1400 with p-value 0.0120. Fixing other covariates, a male patient who is older than 60 years old has 34.86% of the risk a male younger than 60 years old has of succumbing to the disease, where

$$\frac{\exp(-2.113 \times 0 - 1.0538 \times 1 + 2.14 \times 0)}{\exp(-2.113 \times 0 - 1.0538 \times 0 + 2.14 \times 0)} = \exp(-1.0538) = .3486.$$

Whereas, fixing other covariates, a female patient who is older than 60 years old has 2.963 times the risk a female younger than 60 years old has of succumbing to the disease, where

$$\frac{\exp(-2.113 \times 1 - 1.0538 \times 1 + 2.14 \times 1)}{\exp(-2.113 \times 1 - 1.0538 \times 0 + 2.14 \times 0)} = \exp(1.0862) = 2.963.$$

In Figure 5.2, we plot the interaction between SEX and AGE60 based on the means computed using the S function `survfit` for the response and AGE60, fixing female and male separately. It shows a clear pattern of interaction, which supports the prior numeric results using Cox model cns2.coxint6. In Figure 5.3, we first fit the data to the model

```
> cox.fit <- coxph(Surv(B3TODEATH,STATUS)~ KPS.PRE.+GROUP+SEX
                   +strata(AGE60))
```

We then set GROUP=1, KPS.PRE.=80 for female and male separately and obtain the summary of the quantiles using `survfit` as follows:

```
> cns.fit1.1 <- survfit(cox.fit,data.frame(GROUP=1,SEX=1,
                        KPS.PRE.=80)) # Female
> cns.fit0.1 <- survfit(cox.fit,data.frame(GROUP=1,SEX=0,
                        KPS.PRE.=80)) # Male
```

Figure 5.3 displays ordinal interaction between SEX and AGE60 for the three quartiles.

If one sets the covariate KPS.PRE. equal to different values, one can study its relationship to the interaction as well as its effect on the various estimated quantiles of the survival distribution. However, this is tedious. The "censored regression quantiles" approach introduced by Portnoy (2002) enables one to study each of the estimated quantiles as a function of the targeted covariates. This nonparametric methodology is presented in Chapter 8 of this book.

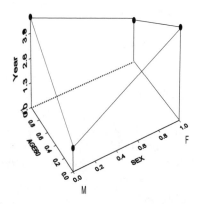

Figure 5.2 *Interaction between* SEX *and* AGE60.

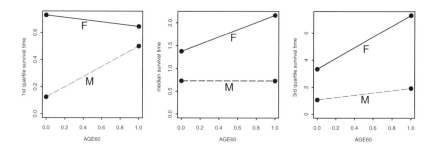

Figure 5.3 *Interaction between* SEX *and* AGE60 *adjusted for* KPS.PRE., GROUP, SEX *via* coxph *and then evaluated at* GROUP = *1 and* KPS.PRE. = *80.*

5.2 Stratified Cox PH regression

We stratify on a categorical variable such as group, gender, and exposure still fitting the other covariates. We do this to obtain nonparametric estimated survival curves for the different levels having adjusted for the other covariates. We then plot the curves to view the estimate of the categorical effect, after adjusting for the effects of the other covariates. If the curves cross or are nonproportional, this implies the existence of the interaction effect unexplained in the model. Then look for appropriate interaction term(s) to include in the model, or stay with the stratified model. If the curves are proportional, this indicates that the interaction effect is well explained by the model you have identified and it supports the Cox PH model. Then use the Cox PH model without the stratification. The disadvantage when we stratify, and the PH assumption is satisfied, is that we cannot obtain an estimated coefficient of the categorical variable effect.

We now apply this procedure to our final model for CNS data. In the following S program we first stratify on the GROUP variable still fitting KPS.PRE., SEX, AGE60, and SEX:AGE60 as covariates. Next, we repeat this procedure for SEX. Again, the disadvantage here is that we cannot obtain an estimated coefficient of the group and sex effects, respectively.

```
> attach(cns2)
> cns2.coxint7 <- coxph(Surv(B3TODEATH,STATUS)~strata(GROUP)
      +KPS.PRE.+SEX+AGE60+SEX:AGE60)
> cns2.coxint7
              coef exp(coef) se(coef)     z      p
 KPS.PRE. -0.0326     0.968   0.0108 -3.03 0.0025
      SEX -2.2028     0.110   0.7195 -3.06 0.0022
    AGE60 -1.1278     0.324   0.4778 -2.36 0.0180
SEX:AGE60  2.2576     9.560   0.8785  2.57 0.0100
```

Likelihood ratio test=20.3 on 4 df, p=0.000433 n= 58

```
> cns2.coxint8 <- coxph(Surv(B3TODEATH,STATUS)~strata(SEX)
     +KPS.PRE.+GROUP+AGE60+SEX:AGE60)
> cns2.coxint8
            coef exp(coef) se(coef)     z      p
KPS.PRE. -0.033    0.968    0.0104 -3.19 0.0014
   GROUP  1.178    3.247    0.3829  3.08 0.0021
   AGE60 -0.994    0.370    0.4552 -2.18 0.0290
SEX:AGE60 2.244    9.427    0.8791  2.55 0.0110
```

Likelihood ratio test=27 on 4 df, p=0.0000199 n= 58

```
# The following gives plots of survival curves resulting from
# stratified Cox PH models to detect any pattern.
# Figure 5.4: upper part.
> par(mfrow=c(2,2))
> survfit.int7 <- survfit(cns2.coxint7)
> plot(survfit.int7,col=1,lty=3:4,lwd=2,cex=3,label=c(10,10,7),
    xlab="Survival time in years from first BBBD",
       ylab="Percent alive",yscale=100)
> legend(3.0,.92,c("group=0","group=1"),lty=3:4,lwd=2)
> survfit.int8 <- survfit(cns2.coxint8)
> plot(survfit.int8,col=1,lty=3:4,lwd=2,cex=3,label=c(10,10,7),
    xlab="Survival time in years from first BBBD",
       ylab="Percent alive",yscale=100)
> legend(3.8,.6,c("male","female"),lty=3:4,lwd=2)
```

For the Weibull regression model, recall (4.5) the log of the cumulative hazard function is linear in $\log(t)$. In general, when we look at the Cox PH model as well as the Weibull model, the plot of $H(t)$ against t on a log-log scale can be very informative. In the `plot` function, the optional function "fun="cloglog"" takes the `survfit` object and plots $H(t)$ against t on a log-log scale.

The following S code plots cumulative hazard functions against t, on a log-log scale, resulting from stratified Cox PH models to detect a nonproportional hazards trend for the SEX and GROUP variables.

```
# Figure 5.4: lower part.
> plot(survfit.int7,fun="cloglog",col=1,lty=3:4,label=c(10,10,7),
    lwd=2,xlab="time in years from first BBBD",
       ylab="log-log cumulative hazard")
> legend(0.05,.8,c("group=0","group=1"),lwd=2,lty=3:4)
> plot(survfit.int8,fun="cloglog",col=1,lty=3:4,label=c(10,10,7),
```

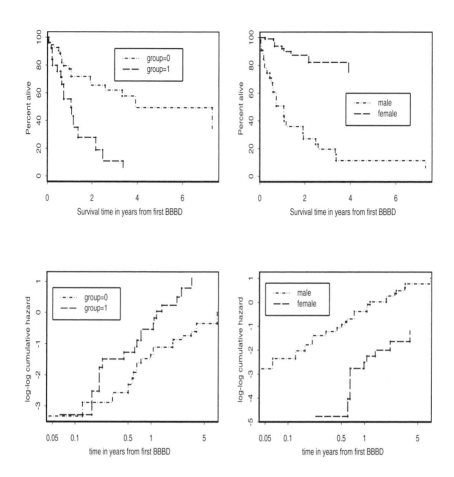

Figure 5.4 *Stratified survivor and log-log cumulative hazards plots to check for PH assumption.*

```
   lwd=2,xlab="time in years from first BBBD",
     ylab="log-log cumulative hazard")
> legend(0.05,.8,c("male","female"),lwd=2,lty=3:4)
> detach()
```
Discussion

- Figure 5.4 shows clear differences between the two groups and between the males and females, respectively. Further, for both GROUP and SEX, the two curves are proportional. This supports the Cox PH model.

- Stratification doesn't change the p-values of the variables in the model cns2.coxint6. The estimated coefficients are very close as well. That is, the model cns2.coxint6 explains all the interaction among the covariates.

Remarks: The Cox PH model formula says that the hazard at time t is the product of two quantities $h_0(t)$, an unspecified baseline hazard function, and $\exp(\sum_{j=1}^{m} \beta_j x^{(j)})$. The key features of the PH assumption are that

1. $h_0(t)$ is a function of t, but does not involve the covariates $x^{(j)}$.
2. $\exp(\sum_{j=1}^{m} \beta_j x^{(j)})$ involves the covariates $x^{(j)}$, but does not involve t.

These two key features imply the HR must then be constant with respect to time t. We now provide an example of a situation where the PH assumption is violated.

Example: Extracted from Kleinbaum (1996, pages 109 − 111).

A study in which cancer patients are randomized to either surgery or radiation therapy without surgery is considered. We have a $(0, 1)$ exposure variable E denoting surgery status, with 0 if a patient receives surgery and 1 if not (i.e., receives radiation). Suppose further that this exposure variable is the only variable of interest.

Is the Cox PH model appropriate? To answer this note that when a patient undergoes serious surgery, as when removing a cancerous tumor, there is usually a high risk for complications from surgery or perhaps even death early in the recovery process, and once the patient gets past this early critical period, the benefits of surgery, if any, can be observed.

Thus, in a study that compares surgery to no surgery, we might expect to see hazard functions for each group that appear in Figure 5.5. Notice that these two functions cross at about three days, and that prior to three days the hazard for the surgery group is higher than the hazard for the no surgery group. Whereas, after three days, we have the reverse. For example, looking at the graph more closely, we can see that at two days, when $t = 2$, the HR of no surgery $(E = 1)$ to surgery $(E = 0)$ patients yields a value less than one. In contrast, at $t = 5$ days, the HR is greater than one. Thus, if the description of the hazard function for each group is accurate, the hazard ratio is not constant over time as HR is some number less than one before three days and greater than one after three days. Hence, the PH assumption is violated as the HR does vary with time. **The general rule is that if the hazard functions cross over time, the PH assumption is violated.** If the Cox PH model is inappropriate, there are several options available for the analysis:

- analyze by **stratifying** on the exposure variable; that is, do not fit any regression model, and, instead obtain the Kaplan-Meier curve for each group separately;

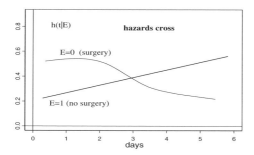

Figure 5.5 *Hazards crossing over time.*

- start the analysis at three days, and use a Cox PH model on three-day survivors;

- fit a Cox PH model for less than three days and a different Cox PH model for greater than three days to get two different hazard ratio estimates, one for each of these two time periods;

- fit a Cox PH model that includes a time-dependent variable which measures the interaction of exposure with time. This model is called an **extended Cox model** and is presented in Chapter 7.1.

- use the **censored regression quantile** approach presented in Chapter 8 allowing crossover effects. This approach is nonparametric and is free of the PH assumption for its validity.

The curious reader might jump ahead to Chapter 7.1 or Chapter 8. In Chapter 7.1 we present an example that explores an epidemiologic study on the treatment of heroin addicts. To model and compare the retention times of two clinics which differ strongly in their overall treatment policies were the primary goals of the study. The PH assumption is violated for the primary exposure variable of interest, clinic. There an extended Cox model is implemented to accommodate this kind of time dependency. In Chapter 8 the Cox model is compared with the regression quantile approach allowing crossover effects.

5.3 Exercises

A. *Applications*

5.1 We work with the diabetes data again. Refer to Exercise 2.3. Instead of a stratified analysis, we will now fit a Cox proportional hazards model:

$$h(t; x^{(1)}, x^{(2)}, x^{(3)}) = h_0(t) \cdot e^{\beta_1 x^{(1)} + \beta_2 x^{(2)} + \beta_3 x^{(3)}} \quad \text{where}$$

$$x^{(1)} = \begin{cases} 0 & \text{man} \\ 1 & \text{woman} \end{cases}$$

$$x^{(2)} = \begin{cases} 0 & \text{nondiabetic} \\ 1 & \text{diabetic} \end{cases}$$

$$x^{(3)} = \text{age in years}$$

(a) Describe the results of the analysis with all three covariates $x^{(1)}$, $x^{(2)}$, and $x^{(3)}$. Which covariates are significant?

Tips:

```
> diab.cox <- coxph(Surv(lzeit,tod) ~ sex+diab+alter,
                    diabetes)
> summary(diab.cox)
```

(b) Stay with estimated full model in part (a) regardless of whether or not the coefficients of the covariates are statistically significantly different from zero.

   i. Use the hazard ratio to estimate the gender effect when the other covariates are held constant. Put woman in the numerator. Interpret!

   ii. Use the hazard ratio to estimate the effect of $x^{(2)}$ and age together for the same gender. Take $x^{(2)} = 1$ and $x^{(3)} = \text{age} + 1$ in the numerator and $x^{(2)} = 0$ and $x^{(3)} = \text{age}$ in the denominator. Interpret!

(c) Simplify (reduce) the model in part (a) as far as possible. Comment!

(d) We now add two additional covariates, which model in the possible dependence of diabetic or not with age. That is, we replace $x^{(3)}$ with the following <u>interaction</u> variables:

$$x^{(4)} = \begin{cases} \text{age, if diabetic} \\ 0 \quad \text{otherwise} \end{cases}$$

$$x^{(5)} = \begin{cases} \text{age, if nondiabetic} \\ 0 \quad \text{otherwise} \end{cases}$$

Now do the analysis as in part (a).

Tips:

```
> diabetes$x4 <- diabetes$x5 <- diabetes$alter
> diabetes$x4[diabetes$diab==0] <- 0
> diabetes$x5[diabetes$diab==1] <- 0
```

(e) Simplify the model in part (d) as much as possible. Draw conclusions (as much as possible as you are not diabetes specialists, etc.) and compare these results with those from your results in the stratified analysis you performed in Exercise 2.4(c).

(f) Use `stepAIC` to select the best subset of variables from the four variables in part (d). Consider only main effects models. Do "backward" elimination starting with all four variables in the model. Refer to Remark 2 on page 129.

(g) Fit your selected model from `stepAIC` to a Cox regression model stratified on the `diab` variable. See page 133.

(h) For the stratified fit in part (g), produce a plot of the survival curves and produce a plot of the log-log cumulative hazards. See pages 134 and 135.

5.2 Replicate Figure 5.3 for GROUP=0 and KPS.PRE.=80. Compare your panel plot to Figure 5.3 and comment.

Tip:

If you name your `survfit` object cns.fit1.0 and cns.fit0.0, then type

```
> summary(cns.fit1.0)
> summary(cns.fit0.0)
```

5.4 Review of first five chapters: self-evaluation

Evaluate your understanding of the basic concepts and methods covered in Chapter 1 through Chapter 5.

1. The data given below are survival times T_i (in months) after an operation:

$$8, \ 1^+, \ 7, \ 7, \ 4^+, \ 5, \ 7^+, \ 5, \ 6^+, 3$$

Suppose that the T_i are iid $\sim$ exponential with p.d.f. $f(x) = \frac{1}{\theta} \exp(-x/\theta)$.

(a) Estimate the mean survival time. That is, compute the MLE of θ.

(b) Construct a 95% confidence interval for the true mean survival time θ.

(c) Report the estimated probability that a patient survives at least half a year.

2. Consider the data above in Problem 1.

(a) Calculate by hand the Kaplan-Meier estimate of survival $\widehat{S}(t)$ (without the standard errors).

(b) Calculate the standard errors s.e.$(\widehat{S}(t))$ at the times $t = 3, \ 4^+$, and 5.

(c) Sketch the Kaplan-Meier curve and mark the censored data.

(d) According to the Kaplan-Meier estimate, how large is the probability that a patient survives at least half a year?

3. In this problem we use data from an ovarian cancer study. The data set is found in the S data.frame `ovarian`. Use S code > `attach(ovarian)` to access this data set. It is also found in R. Use R code > `library(survival)` followed by > `data(ovarian)`. But this is not necessary to answer the problems.

futime	survival time in days after diagnosis of the cancer
fustat	$0 =$ censored, $1 =$ dead
age	age in years
residual.dz	a measure of the health condition after chemotherapy
rx	indicator variable for type of chemo treatment: $1 =$ treatment A, $2 =$ treatment B
ecog.ps	a measure of functioning of the ovaries

Source: Edmunson, J. H., Fleming, T. R., Decker, D. G., Malkasian, G. D., Jefferies, J. A., Webb, M. J., and Kvols, L. K. (1979). Different chemotherapeutic sensitivities and host factors affecting prognosis in advanced ovarian carcinoma vs. minimal residual disease. *Cancer Treatment Reports* **63**, 241–47.

Do the two treatments have different effects on survival?
Justify your answer with the help of the following S output.

```
Call: survdiff(formula=Surv(futime,fustat)~rx,data=ovarian)

        N Observed Expected (O-E)^2/E (O-E)^2/V
rx=1 13        7     5.23     0.596      1.06
rx=2 13        5     6.77     0.461      1.06

Chisq= 1.1  on 1 degrees of freedom, p= 0.303
```

4. Consider the Cox proportional hazards model for the ovarian cancer data in Problem 3.

```
Call: coxph(formula=Surv(futime,fustat)~age+residual.dz+
                    rx+ecog.ps,data=ovarian)
n= 26
               coef  exp(coef)  se(coef)      z       p
        age   0.125      1.133    0.0469   2.662  0.0078
residual.dz   0.826      2.285    0.7896   1.046  0.3000
         rx  -0.914      0.401    0.6533  -1.400  0.1600
    ecog.ps   0.336      1.400    0.6439   0.522  0.6000

Rsquare= 0.481 (max possible= 0.932 )
Likelihood ratio test= 17 on 4 df, p=0.0019
Wald test = 14.2 on 4 df, p=0.00654
Efficient score test = 20.8 on 4 df, p=0.000345
```

As always, justify your answers.

(a) Which variables are significant?

(b) The LRT value from just fitting age is 14.3 with 1 degree of freedom. Perform a LRT type test to test if the full model with the four covariates significantly improves the fit over the model with just age. No need to access the data and use S. The information given here is enough to calculate the value of the appropriate LRT statistic. Then compute the p-value using S or EXCEL, etc.

(c) Continue to work with the full model. How does age effect survival time? Does survival increase or decrease with age? Justify by the hazard function and the HR as a measure of effect.

(d) Suppose rx is significant.
Which treatment gives the larger survival probability? Again, be sure to use the HR to quantify the effects.

5. We continue to work with the ovarian cancer data. Instead of fitting a Cox PH model, we fit the data to a Weibull regression model:

```
survReg(Surv(futime,fustat) ~ age+residual.dz+rx+ecog.ps,
        data=ovarian, dist="weib")

# Model after backward elimination (stepwise back,
# one-variable at a time reduction):

Call: survReg(formula=Surv(futime,fustat) ~ age,
              data=ovarian,distribution="weibull")

Coefficients:
             Est. Std.Err. 95% LCL  95% UCL  z-value   p-value
(Intercept) 12.4   1.482    9.492   15.302     8.36   6.05e-017
        age -0.1   0.024   -0.143    -0.05    -4.06   4.88e-005

Extreme value distribution: Dispersion (scale) = 0.6114563
Observations: 26 Total; 14 Censored -2*Log-Likelihood: 180
```

(a) Specify the estimated hazard function.

(b) How does age effect survival time? Does survival increase or decrease with age?

(c) How many years after diagnosis will 50% of the patients with age 40 years die?

CHAPTER 6

Model Checking: Data Diagnostics

Objectives of this chapter:

After studying Chapter 6, the student should:

1. Know and understand the definition of **model deviance**:

   (a) likelihood of fitted model

   (b) likelihood of saturated model

   (c) deviance residual.

2. Be familiar with the term **hierarchical models**.

3. Know the definition of **partial deviance**, its relationship to the likelihood ratio test statistic, and how we use it to reduce models and test for overall model adequacy.

4. Know how to interpret the measure **dfbeta**.

5. Know that the S function `survReg` along with companion function `resid` provides the deviance residuals, `dfbeta`, and `dfbetas`.

6. Be familiar with Cox's **partial likelihood** function.

7. Be familiar with and how to use the following residuals to assess the various proportional hazards model assumptions:

   (a) *Cox-Snell residuals*

   (b) *Martingale residuals*

   (c) *Deviance residuals*

   (d) *Schoenfeld residuals*

   (e) *Scaled Schoenfeld residuals*

   (f) *dfbetas.*

8. Be familiar with the S functions `coxph` and `cox.zph` and which residuals these functions provide.

9. Know the definition of **profile likelihood function** and be able to conduct a *cut point analysis with bootstrap validation*.

6.1 Basic graphical methods

When searching for a parametric model that fits the data well, we use graphical displays to check the model's appropriateness; that is, the goodness of fit. Miller (1981, page 164) points out that "the human eye can distinguish well between a straight line and a curve." We quote Miller's basic principle as it should guide the method of plotting.

Basic principle:

Select the scales of the coordinate axes so that if the model holds, a plot of the data resembles a straight line, and if the model fails, a plot resembles a curve.

The construction of the Q-Q plot (page 63) for those log-transformed distributions, which are members of the location and scale family of distributions, follows this basic principle. The linear relationships summarized in Table 3.1, page 64, guided this construction. Some authors, including Miller, prefer to plot the uncensored points (y_i, z_i), $i = 1, \cdots, r \leq n$. This plot is commonly called a **probability plot**. We prefer the convention of placing the log data y_i on the vertical axis and the standard quantiles z_i on the horizontal axis; hence, the Q-Q plot.

The S function `survReg` only fits models for log-time distributions belonging to the location and scale family. For this reason we have ignored the gamma model until now. A Q-Q plot is still an effective graphical device for non-members of the location and scale family. For these cases, we plot the ordered uncensored times t_i against the corresponding quantiles q_i from the distribution of interest. If the model is appropriate, the points should lie very close to the 45°-line through the origin $(0, 0)$. We compute and plot the quantiles based on the K-M estimates against the quantiles based on the parametric assumptions. That is, for each uncensored t_i, compute $\widehat{p}_i = 1 - \widehat{S}(t_i)$, where $\widehat{S}(t_i)$ denotes the K-M estimate of survival probability at time t_i. Then, with this set of probabilities, compute the corresponding quantiles q_i from the assumed distribution with MLE's used for the parameter values. Finally, plot the pairs (q_i, t_i). Note that $\widehat{p}_i = 1 - \widehat{S}(t_i) = 1 - \widehat{S}_{\text{model}}(q_i)$. To compute the MLE's for the unknown parameters in S, the two functions available are `nlmin` and `nlminb`. As these functions find a local minimum, we use these functions to minimize $(-1) \times$ the log-likelihood function. For our example, we draw the Q-Q plot for the AML data fit to a gamma model. In this problem, we must use `nlminb` since the gamma has bound-constrained parameters; that is, $k > 0$ and $\lambda > 0$, corresponding to shape and scale, respectively. The function `qq.gamma` gives the Q-Q plot for data fit to a gamma. See Figure 6.1.

```
> attach(aml)
 # Q-Q plot for maintained group
> weeks.1 <- weeks[group==1]
> status.1 <- status[group==1]
```

```
> weeks1 <- list(weeks.1)
> status1 <- list(status.1)
> qq.gamma(Surv(weeks.1,status.1),weeks1,status1)
 # The 2nd and 3rd arguments must be list objects.
    shape      rate
 1.268666 0.0223737  #MLE's
 # Q-Q plot for nonmaintained group
> weeks.0 <- weeks[group == 0]
> status.0 <- status[group == 0]
> weeks0 <- list(weeks.0)
> status0 <- list(status.0)
> qq.gamma(Surv(weeks.0,status.0),weeks0,status0)
    shape      rate
 1.987217 0.08799075  # MLE'S
> detach()
```

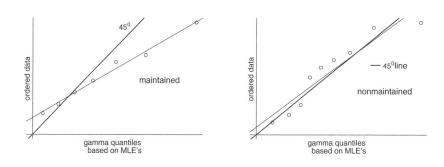

Figure 6.1 *Q-Q plot for AML data fit to gamma model. MLE's used for parameter values. Points are fit to least squares line.*

It's important to draw the 45°-line. For without the comparison, the least squares line fitted only to uncensored times would have led us to believe the gamma model fit the maintained group well. But this is quite the contrary. The fit is very poor in the upper tail. The estimated gamma quantiles q_i are markedly larger than their corresponding sample quantiles t_i. One reason for this over-fit is the MLE's are greatly influenced by the presence of the one extreme value 161+. It is clear from the previous Weibull, log-logistic, and log-normal Q-Q plots (Figure 3.12, page 87), the log-logistic is a much better choice to model the AML maintained group. Notice the gamma Q-Q plot for this group has a similar pattern to the Weibull Q-Q plot. In contrast, the gamma seems to fit the nonmaintained group quite well. There are no extreme values in this group.

For the two sample problem, let $x = 1$ and $x = 0$ represent the two groups. To check the validity of the Cox PH model, recall from Chapter 4.3 that

$h(t|1) = \exp(\beta)h(t|0)$, where $\exp(\beta)$ is constant with respect to time. This implies $S(t|1) = (S(t|0))^{\exp(\beta)}$ or $\log S(t|1) = \exp(\beta)\log S(t|0)$. Hence, the plots of the ratios are horizonal lines. These graphs are displayed in Figure 6.2. The plots of the empirical quantities constructed with the K-M estimate for

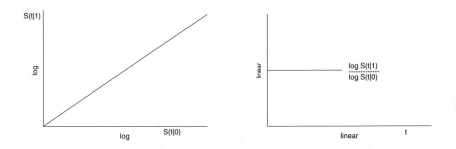

Figure 6.2 *Graph of cumulative hazards ratio.*

each group should reflect the foregoing relationships if the PH assumption is satisfied.

Equivalently, we can plot the empirical hazards for each group on the same plot. The curves should be approximately parallel over time to validate the PH assumption. See Figure 2.5, page 46. It is clear the AML data violate the PH assumption.

To check for a shift by translation, calculate the K-M estimate of survival for each group separately and plot. The curves should be vertically parallel. For example, as the log-gamma is a location family, this plot is useful. An example is displayed in Figure 6.3.

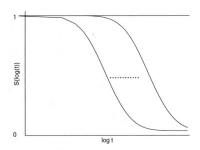

Figure 6.3 *A graph to check for a shift by translation.*

6.2 Weibull regression model

In this section we continue to work with the Motorette data first presented and analyzed in Chapter 4.6, page 107. There AIC selects the Weibull model as the best model and the Q-Q plot supports this. We now consider model diagnostics. We delay the S code until all relevant new definitions are presented.

Recall from expressions (4.1) and (4.4) the Weibull regression model has hazard and survivor functions

$$h(t|\underline{x}) = h_0(t) \cdot \exp(\underline{x}'\underline{\beta}) = \alpha \cdot (\widetilde{\lambda})^\alpha \cdot t^{\alpha-1}, \quad \text{where} \quad \widetilde{\lambda} = \lambda \cdot \left(\exp(\underline{x}'\underline{\beta})\right)^{\frac{1}{\alpha}},$$

and

$$S(t|\underline{x}) = \exp\left(-(\widetilde{\lambda}t)^\alpha\right).$$

The log of the cumulative hazard (4.5) is

$$\log\left(H(t|\underline{x})\right) = \log\left(-\log\left(S(t|\underline{x})\right)\right) = \alpha\log(\lambda) + \underline{x}'\underline{\beta} + \alpha\log(t).$$

Expression (4.3) tells us

$$Y = \log(t) = \underline{x}'\underline{\beta}^* + \beta_0^* + \sigma \cdot Z,$$

where $Z \sim$ standard extreme value distribution.

Graphical checks of overall model adequacy

We see that $\log(t)$ is not only linear in z, but also in each $x^{(j)}$, $j = 1, \ldots, m$. For exploring whether or not the Weibull distribution is a suitable model for the data at the (ordered) uncensored times t_i, we draw a Q-Q plot (page 63) of the points (z_i, y_i), and we draw m plots of the points $(x_i^{(j)}, y_i)$, $i = 1, \cdots, r \le n$ and $j = 1, \cdots, m$. To support the Weibull regression model, all plots should display straight lines. See Figures 6.4 and 6.5. If not, perhaps transforming those particular $x^{(j)}$'s and/or t_i could improve the fit. If not, try another model.

The Q-Q plot is also very useful for detecting overall adequacy of the final reduced regression model; that is, goodness-of-fit. As the single covariate x in the Motorette data has three distinct levels, we draw two Q-Q plots. In Figure 6.8, each group is fit to its own Weibull. The lines have different slopes and intercepts. In Figure 6.9, we fit a regression model with covariate x. The lines have same slope, but different intercepts. These plots can reveal additional information masked in Figures 6.4 and 6.5.

The `survReg` procedure in S gives the MLE's

$$\widehat{\beta_0^*}, \quad \widehat{\underline{\beta}^*}, \quad \widehat{\sigma}, \quad \text{and} \quad \widehat{\widetilde{\mu}} = \widehat{\beta_0^*} + \underline{x}'\widehat{\underline{\beta}^*}. \tag{6.1}$$

For the Weibull parameters we have

$$\widehat{\lambda} = \exp(-\widehat{\beta_0^*}), \quad \widehat{\underline{\beta}} = -\widehat{\alpha}\,\widehat{\underline{\beta}^*}, \quad \widehat{\alpha} = 1/\widehat{\sigma}, \quad \text{and} \quad \widehat{\widetilde{\lambda}} = \exp(-\widehat{\widetilde{\mu}}). \tag{6.2}$$

Note that `survReg` provides the **fitted times** $\widehat{T}_i$. So,

$$\widehat{Y}_i = \log(\widehat{T}_i) = \widehat{\widetilde{\mu}}_i \ . \tag{6.3}$$

Also recall (page 58) the p.d.f. of $Y_i = \log(T_i)$ and the corresponding survivor function evaluated at these estimates are

$$f(y_i|\widehat{\widetilde{\mu}}_i, \widehat{\sigma}) = \frac{1}{\widehat{\sigma}} \exp\left(\frac{y_i - \widehat{\widetilde{\mu}}_i}{\widehat{\sigma}} - \exp\left(\frac{y_i - \widehat{\widetilde{\mu}}_i}{\widehat{\sigma}}\right)\right) \tag{6.4}$$

$$S(y_i|\widehat{\widetilde{\mu}}_i, \widehat{\sigma}) = \exp\left(-\exp\left(\frac{y_i - \widehat{\widetilde{\mu}}_i}{\widehat{\sigma}}\right)\right). \tag{6.5}$$

Deviance, deviance residual, and graphical checks for outliers

We now consider a measure useful in detecting outliers. Define the **model deviance** as

$$\mathcal{D} = -2 \times (\text{log-likelihood of the fitted model} - \text{log-likelihood of}$$
$$\text{the saturated model})$$

$$= -2 \times \left(\sum_{i=1}^{n}\left(\log(\widehat{L}_i) - \log(\widehat{L}_{si})\right)\right), \tag{6.6}$$

where $\widehat{L}_i$ denotes the ith individual's likelihood evaluated at the MLE's, and $\widehat{L}_{si}$ denotes the ith factor in the saturated likelihood evaluated at the MLE of θ_i. A saturated model is one with n parameters that fit the n observations perfectly. Let $\theta_1, \ldots, \theta_n$ denote the n parameters. This also entails that we obtain these MLE's with **no** constraints. According to Klein & Moeschberger (1997, page 359), in computing the deviance the nuisance parameters are held fixed between the fitted and the saturated model. In the Weibull regression model, the only nuisance parameter is the σ and is held fixed at the MLE value obtained in the fitted model. The measure $\mathcal{D}$ can be used as a goodness of fit criterion. The larger the model deviance, the poorer the fit and vice versa. For an approximate size-α test, compare the calculated $\mathcal{D}$ value to the χ_α^2 critical value with $n - m - 1$ degrees of freedom.

Under the random (right) censoring model and under the assumption that censoring time has no connection with the survival time, recall the **likelihood function** of the sample (1.13) is

$$L = L(\beta_0^*; \underline{\beta}^*; \sigma) = L(\underline{\mu}; \sigma) = \prod_{i=1}^{n} L_i(\widetilde{\mu}_i; \sigma),$$

where

$$L_i(\widetilde{\mu}_i; \sigma) = \left(f(y_i|\widetilde{\mu}_i, \sigma)\right)^{\delta_i} \left(S(y_i|\widetilde{\mu}_i, \sigma)\right)^{1-\delta_i} \quad \text{and}$$

$$\delta_i = \begin{cases} 1 & \text{if } y_i \text{ is uncensored} \\ 0 & \text{if } y_i \text{ is censored.} \end{cases}$$

In preparation to define the deviance residual, we first define two types of residuals which are the parametric analogues to those defined and discussed in some detail in Section 6.3.

Cox-Snell residual

The ith *Cox-Snell residual* is defined as

$$r_{C\,i} = \widehat{H}_0(t_i) \times \exp(\underline{x}_i'\widehat{\underline{\beta}}), \tag{6.7}$$

where $\widehat{H}_0(t_i)$ and $\widehat{\underline{\beta}}$ are the MLE's of the baseline cumulative hazard function and coefficient vector, respectively. As these residuals are always nonnegative, their plot is difficult to interpret. These are not residuals in the sense of linear models because they are not the difference between the observed and fitted values. Their interpretation is discussed in Section 6.3.

Martingale residual

The ith *martingale residual* is defined as

$$\widehat{M}_i = \delta_i - r_{C\,i}. \tag{6.8}$$

The $\widehat{M}_i$ take values in $(-\infty, 1]$ and are always negative for censored observations. In large samples, the martingale residuals are uncorrelated and have expected value equal to zero. But they are not symmetrically distributed about zero.

Deviance residual

The ith *deviance residual*, denoted by D_i, is the square root of the ith term of the deviance, augmented by the sign of the $\widehat{M}_i$:

$$D_i = \text{sign}(\widehat{M}_i) \times \sqrt{-2 \times \left(\log\left(\widehat{L}_i(\widehat{\underline{\mu}}_i, \widehat{\sigma})\right) - \log(\widehat{L}_{si})\right)}. \tag{6.9}$$

These residuals are expected to be symmetrically distributed about zero. Hence, their plot is easier to interpret. But we caution these do not necessarily sum to zero. The model deviance then is

$$\mathcal{D} = \sum_{i=1}^{n} D_i{}^2 = \text{the sum of the squared deviance residuals.}$$

When there is light to moderate censoring, the D_i should look like an iid normal sample. Therefore, the deviance residuals are useful in detecting outliers. To obtain the D_i, use > `resid(fit,type="deviance")` where `fit` is a `survReg` object. A plot of the D_i against the fitted log-times is given in Figure 6.6.

There are three plots constructed with D_i that are very useful in helping to detect outliers. One is the normal probability plot. Here we plot the kth ordered D_i against its normal score $Z((k-.375)/(n+.25))$ where $Z(A)$ denotes the Ath quantile of the standard normal distribution. Outliers will be points that fall substantially away from a straight line. The second graph plots the D_i against the estimated <u>risk scores</u> $\sum_{j=1}^{m} \widehat{\beta}_j^* x_i^{(j)}$. This plot should look like a scatter of random noise about zero. Outliers will have large absolute deviations and will sit apart from the point cloud. The third graph plots D_i against its observation (index) number. Again, we look for points that are set apart with large absolute value. See Figure 6.10.

For the interested reader, we give the expressions for the **ith deviance residual (6.9) under the extreme value model**. The ith factor in the saturated likelihood is

$$\widehat{L}_{si}(y_i; \widehat{\theta}_i, \widehat{\sigma}) = \left(\frac{e^{-1}}{\widehat{\sigma}}\right)^{\delta_i}. \tag{6.10}$$

WHY! Then for an uncensored observation ($\delta_i = 1$), the ith deviance residual is

$$\begin{aligned}
D_i &= \text{sign}(\widehat{M}_i) \times \sqrt{-2 \times \left(\log\left(\widehat{L}_i(\widehat{\widetilde{\mu}}_i, \widehat{\sigma})\right) - \log(\widehat{L}_{si})\right)} \\
&= \text{sign}(\widehat{M}_i) \times \sqrt{2 \times \left(-\left(\frac{y_i - \widehat{\widetilde{\mu}}_i}{\widehat{\sigma}}\right) + \exp\left(\frac{y_i - \widehat{\widetilde{\mu}}_i}{\widehat{\sigma}}\right) - 1\right)}. \tag{6.11}
\end{aligned}$$

For a censored observation ($\delta_i = 0$), as the $\widehat{M}_i$ are always negative in this case,

$$D_i = -\sqrt{w_i \times 2 \times \left(\exp\left(\frac{y_i - \widehat{\widetilde{\mu}}_i}{\widehat{\sigma}}\right)\right)}, \tag{6.12}$$

where the weight w_i equals the number of observations of same censored time value and same values for the covariates. Data are often recorded in spreadsheets with a column of such weights. The weight is equal to 1 for uncensored points.

For the Weibull case, the sign of $\widehat{M}_i$ in the D_i is motivated as follows:

1. It follows from expressions (6.1) and (6.2)

$$\frac{y_i - \widehat{\widetilde{\mu}}_i}{\widehat{\sigma}} = \log(\widehat{\lambda}t_i)^{\widehat{\alpha}} + \underline{x}_i'\widehat{\underline{\beta}}. \tag{6.13}$$

Therefore, it follows from expressions (6.13) and (4.6)

$$\exp\left(\frac{y_i - \widehat{\widetilde{\mu}}_i}{\widehat{\sigma}}\right) = (\widehat{\lambda}t_i)^{\widehat{\alpha}} \times \exp(\underline{x}_i'\widehat{\underline{\beta}}) = r_{Ci}. \tag{6.14}$$

WHY!

2. When $\delta_i = 0$,

$$D_i^2 = -2 \times (-r_{Ci}) > 0. \tag{6.15}$$

When $\delta_i = 1$,
$$D_i^2 = -2 \times \left(1 - r_{C\,i} + \log(r_{C\,i})\right) \geq 0. \tag{6.16}$$

WHY!

Collett (1994, page 154) views the deviance residuals as martingale residuals that have been transformed to produce values that are symmetric about zero when the fitted model is appropriate. Hence, we match the sign of deviance residuals with the sign of corresponding martingale residuals. In light of expressions (6.15) and (6.16), we can express the deviance residual for Weibull regression as

$$D_i = \text{sign}(\widehat{M_i}) \times \sqrt{-2 \times \left\{\widehat{M_i} + \delta_i \log(\delta_i - \widehat{M_i})\right\}}. \tag{6.17}$$

This now matches the definition of deviance residual to be presented in Section 6.3.3.

We note that the definition stated in Klein & Moeschberger (page 396) always assigns a "negative" sign to the deviance residuals corresponding to censored times. S does just the opposite. A "plus" is always assigned. Klein & Moeschberger are correct. The following should explain what S does: When $\delta_i = 1$, it is easy to show that

$$\text{sign}(\widehat{M_i}) = \text{sign}(1 - r_{C\,i}) = -\text{sign}\left(\log(r_{C\,i})\right) = -\text{sign}(y_i - \widehat{\bar{\mu}}_i). \tag{6.18}$$

WHY! Hence, for graphical purposes, if no censoring were present, it would be equivalent to using the $\text{sign}(y_i - \widehat{\bar{\mu}}_i)$ as we only look for random scatter about zero.

When $\delta_i = 0$, the $\text{sign}(\widehat{M_i})$ equals the $\text{sign}(-r_{C\,i})$, which is always negative. To be consistent with the definition of deviance residual in the nonlinear regression model with no censoring, S uses the $\text{sign}(y_i - \widehat{\bar{\mu}}_i)$. Thus, they assign a "plus" to D_i corresponding to censored observations and the opposite sign to D_i corresponding to uncensored observations.

Partial deviance

We now consider hierarchical (nested) models. Let R denote the reduced model and F denote the full model which consists of additional covariates added to the reduced model. Partial deviance is a measure useful for model building. We define **partial deviance** as

$$
\begin{aligned}
\mathcal{PD} &= \text{Deviance (additional covariates} \mid \text{covariates in the reduced model)} \\
&= \mathcal{D}(R) - \mathcal{D}(F) = -2\log\left(\widehat{L}(R)\right) + 2\log\left(\widehat{L}(F)\right) \tag{6.19} \\
&= -2\log\left(\frac{\widehat{L}(R)}{\widehat{L}(F)}\right).
\end{aligned}
$$

We see that the partial deviance is equivalent to the LRT statistic. Hence, the

LRT checks to see if there is significant partial deviance. We reject when $\mathcal{PD}$ is "large." If the partial deviance is large, this indicates that the additional covariates improve the fit. If the partial deviance is small, it indicates they don't improve the fit and the smaller model (the reduced one) is just as adequate. Hence, drop the additional covariates and continue with the reduced model. Partial deviance is analogous to the extra sum of squares, SSR(additional covariates|covariates in reduced model), for ordinary linear regression models. In fact, when the $\log(t_i)$'s are normal and no censoring is present, partial deviance simplifies to the corresponding extra sum of squares. WHY!

dfbeta

dfbeta is a useful measure to assess the influence of each point on the estimated coefficients $\widehat{\beta}_j$'s. This measure is analogous to that used in regular linear regression. Large values suggest we inspect corresponding data points. The measure **dfbetas** is dfbeta divided by the s.e.$(\widehat{\beta}_j)$. We obtain these quantities via the companion function `resid` where `fit` is a `survReg` object.
`> resid(fit, type="dfbeta")`.
See Figure 6.7 for a plot of the dfbeta for each observation's influence on the coefficient of the x variable. See Section 6.3.6 for a more detailed discussion of the dfbeta measure.

Motorette example:

Is the Weibull regression model appropriate?

Figure 6.4:

```
> attach(motorette)
> qq.weibull(Surv(time,status))
```

Figure 6.5:

```
> plot.logt.x(time,status,x) # Plot of log(t) against x.

# Now the Weibull regression fit:
> motor.fit <- survReg(Surv(time,status) ~ x,dist="weibull")
> dresid <- resid(motor.fit,type="deviance")
> riskscore <- log(fitted(motor.fit)) - coef(motor.fit)[1]
```

Figure 6.6:

```
> plot(log(fitted(motor.fit)),dresid)
> mtext("Deviance Residuals vs log Fitted Values (muhat)",
        3,-1.5)
> abline(h=0)
```

Figure 6.10:

```
> index <- seq(1:30)
> par(mfrow=c(2,2))
> plot(riskscore,dresid,ylab="deviance residuals")
> abline(h=0)
> qqnorm.default(dresid,datax=F,plot=T,
                    ylab="deviance residuals")
> qqline(dresid)
> plot(index,dresid,ylab="deviance residual")
> abline(h=0)
```

Figure 6.7:

```
 # We plot dfbeta to assess influence of each point on the
 # estimated coefficient.
> dfbeta <- resid(motor.fit,type="dfbeta")
> plot(index,dfbeta[,1],type="h",ylab="Scaled change in
         coefficient",xlab="Observation")
```

Figure 6.8:

```
> xln <- levels(factor(x))
> ts.1 <- Surv(time[as.factor(x)==xln[1]],
                  status[as.factor(x)==xln[1]])
> ts.2 <- Surv(time[as.factor(x)==xln[2]],
                  status[as.factor(x)==xln[2]])
> ts.3 <- Surv(time[as.factor(x)==xln[3]],
                  status[as.factor(x)==xln[3]])
> qq.weibull(list(ts.1,ts.2,ts.3))
```

Figure 6.9:

```
> xln <- levels(factor(x))
> ts.1 <- Surv(time[as.factor(x)==xln[1]],
                  status[as.factor(x)==xln[1]])
> ts.2 <- Surv(time[as.factor(x)==xln[2]],
                  status[as.factor(x)==xln[2]])
> ts.3 <- Surv(time[as.factor(x)==xln[3]],
                  status[as.factor(x)==xln[3]])
> qq.weibreg(list(ts.1,ts.2,ts.3),motor.fit)
```

We compute the log-likelihood of saturated model, partial deviance, and then compare to the output from the anova function.

```
> summary(motor.fit)
              Value Std. Error    z       p
(Intercept) -11.89      1.966 -6.05 1.45e-009
          x   9.04      0.906  9.98 1.94e-023
Log(scale)  -1.02      0.220 -4.63 3.72e-006


Scale= 0.361


Loglik(model)= -144.3   Loglik(intercept only)= -155.7
    Chisq= 22.67 on 1 degrees of freedom, p= 1.9e-006
# Chisq=22.67 is the LRT value for testing the
# significance of the x variable.
> loglikR <- motor.fit$loglik[1]
> loglikR      # Model has only intercept.
[1] -155.6817
> loglikF <- motor.fit$loglik[2]
> loglikF      # Model includes the covariate x.
[1] -144.3449
> ModelDev <- sum(resid(motor.fit,type="deviance")^2)
> ModelDev
[1] 46.5183  # Full model deviance
> loglikSat <- loglikF + ModelDeviance/2
> loglikSat
[1] -121.0858
> nullDev <- - 2*(loglikR - loglikSat)
> nullDev  # Reduced Model (only intercept)
[1] 69.19193
> PartialDev <- nullDev - ModelDev
> PartialDev
[1] 22.67363 # which equals the LRT value.
    # The following ANOVA output provides Deviance
    # which is really the partial deviance. This is
    # easily seen.
> anova(motor.fit)
Analysis of Deviance Table Response: Surv(time,status)
Terms added sequentially (first to last)
      Df  Deviance Resid.  Df   -2*LL       Pr(Chi)
NULL                        2  311.3634
    x   -1  22.67363        3  288.6898    1.919847e-006
> detach()
```

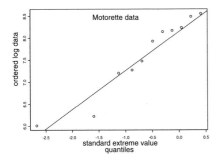

Figure 6.4 *Q-Q plot. MLE's used for slope and intercept.*

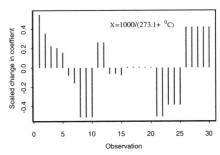

Figure 6.7 *The dfbeta plot helps assess each point's influence on* $\widehat{\beta}$.

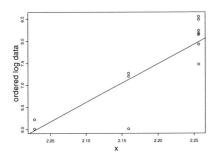

Figure 6.5 *Log(t) against x. Least squares line.*

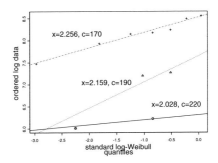

Figure 6.8 *Q-Q plot. Different intercepts and slopes.*

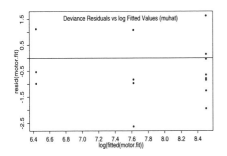

Figure 6.6 *Deviance residuals against fitted log-times.*

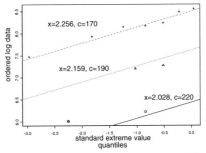

Figure 6.9 *Q-Q plot for model* $y = \beta_0^* + \beta_1^* x +$ *error. Each line based on MLE's. Lines have same slope, but different intercepts.*

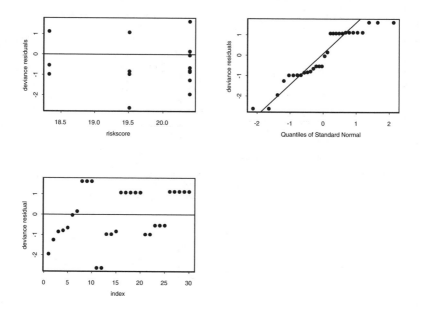

Figure 6.10 *Motorette data: deviance residuals against risk scores, normal quantiles, and index.*

Results:

- In Figure 6.8, each group is fit separately. The graphs suggest the Weibull model gives an adequate description of each group.

- Figure 6.9 supports the Weibull regression model describes well the role temperature plays in the acceleration of failure of the motorettes.

- Figures 6.4 and 6.5 display straight lines. Figure 6.6 displays a random scatter around zero except for a possible outlier whose residual value is -2.634. Figure 6.7 shows there are no influential points. In Figure 6.10, the deviance vs. index plot displays two possible outliers whereas each of the other two plots reveal only one possible outlier.

- The LRT per the `anova` function, with a p-value of 1.9×10^{-6}, provides strong evidence the Weibull model with the predictor variable x is adequate. Equivalently, the p-value of 1.94×10^{-23} for the estimated coefficient of x provides this strong evidence.

6.3 Cox proportional hazards model

Recall from Chapter 4.3 that this model has hazard function

$$
\begin{aligned}
h(t|\underline{x}) = h_0(t) \cdot \exp(\underline{x}'\underline{\beta}) &= h_0(t) \cdot \exp(\beta_1 x^{(1)} + \cdots + \beta_m x^{(m)}) \\
&= h_0(t) \cdot \exp(\beta_1 x^{(1)}) \times \exp(\beta_2 x^{(2)}) \times \cdots \times \exp(\beta_m x^{(m)}),
\end{aligned}
$$

where at two different points $\underline{x}_1, \underline{x}_2$, the proportion

$$
\frac{h(t|\underline{x}_1)}{h(t|\underline{x}_2)} = \frac{\exp(\underline{x}_1'\underline{\beta})}{\exp(\underline{x}_2'\underline{\beta})} = \exp\left((\underline{x}_1' - \underline{x}_2')\underline{\beta}\right),
$$

called the hazards ratio (HR), is constant with respect to time t.

As the baseline hazard function is not specified in the Cox model, the likelihood function cannot be fully specified. To see this, recall that

$$
f(\cdot) = h(\cdot) \times S(\cdot).
$$

The hazard function $h(\cdot)$ depends on the baseline hazard $h_0(\cdot)$. Hence, so does the p.d.f. Cox (1975) defines a likelihood based on conditional probabilities which are free of the baseline hazard. His estimate is obtained from maximizing this likelihood. In this way he avoids having to specify $h_0(\cdot)$ at all. We derive this likelihood heuristically. Let t^* denote a time at which a death has occurred. Let $\mathcal{R}(t^*)$ be the risk set at time t^*; that is, the indices of individuals who are alive and not censored just before t^*. First,

$$
\begin{aligned}
P\{\text{one death in } [t^*, t^* + \triangle t^*) \mid \mathcal{R}(t^*)\} \\
= \sum_{l \in \mathcal{R}(t^*)} P\{T_l \in [t^*, t^* + \triangle t^*) \mid T_l \geq t^*\} \\
\approx \sum_{l \in \mathcal{R}(t^*)} h(t^*|\underline{x}_l)\triangle t^* \\
= \sum_{l \in \mathcal{R}(t^*)} h_0(t^*) \cdot \exp(\underline{x}_l'\underline{\beta})\triangle t^*.
\end{aligned}
$$

Thus, if we let $P\{\text{one death at } t^* \mid \mathcal{R}(t^*)\}$ denote the

$$
\sum_{l \in \mathcal{R}(t^*)} P(T_l = t^*|T_l \geq t^*),
$$

then we have

$$
P\{\text{one death at } t^* \mid \mathcal{R}(t^*)\} = \sum_{l \in \mathcal{R}(t^*)} h_0(t^*) \cdot \exp(\underline{x}_l'\underline{\beta}).
$$

Now, let $t_{(1)}, \ldots, t_{(r)}$ denote the $r \leq n$ distinct ordered (uncensored) death times, so that $t_{(j)}$ is the jth ordered death time. Let $\underline{x}_{(j)}$ denote the vector of covariates associated with the individual who dies at $t_{(j)}$. Then, for each j, we have

$$
L_j(\underline{\beta}) = P\{\text{individual with } \underline{x}_{(j)} \text{ dies at } t_{(j)} \mid \text{one death in } \mathcal{R}(t_{(j)}) \text{ at } t_{(j)}\}
$$

$$= \frac{P\{\text{individual with } \underline{x}_{(j)} \text{ dies at } t_{(j)} \mid \text{individual in } \mathcal{R}(t_{(j)})\}}{P\{\text{one death at } t_{(j)} \mid \mathcal{R}(t_{(j)})\}}$$

$$= \frac{h_0(t_{(j)}) \cdot \exp(\underline{x}'_{(j)}\underline{\beta})}{\sum_{l \in \mathcal{R}(t_{(j)})} h_0(t_{(j)}) \cdot \exp(\underline{x}'_l\underline{\beta})}$$

$$= \frac{\exp(\underline{x}'_{(j)}\underline{\beta})}{\sum_{l \in \mathcal{R}(t_{(j)})} \exp(\underline{x}'_l\underline{\beta})}.$$

The product of these over the r uncensored death times yields what Cox refers to as the partial likelihood. The **partial likelihood function**, denoted by $L_c(\underline{\beta})$, is thus defined to be

$$L_c(\underline{\beta}) = \prod_{j=1}^{r} L_j(\underline{\beta}) = \prod_{j=1}^{r} \frac{\exp(\underline{x}'_{(j)}\underline{\beta})}{\sum_{l \in \mathcal{R}(t_{(j)})} \exp(\underline{x}'_l\underline{\beta})}. \tag{6.20}$$

Recall that in the random censoring model we observe the times $y_1, \ldots, y_n$ along with the associated $\delta_1, \ldots, \delta_n$ where $\delta_i = 1$ if the y_i is uncensored (i.e., the actual death time was observed) and $\delta_i = 0$ if y_i is censored. We can now give an equivalent expression for the partial likelihood function in terms of all n observed times:

$$L_c(\underline{\beta}) = \prod_{i=1}^{n} \left(\frac{\exp(\underline{x}'_i\underline{\beta})}{\sum_{l \in \mathcal{R}(y_i)} \exp(\underline{x}'_l\underline{\beta})} \right)^{\delta_i}. \tag{6.21}$$

Remarks:

1. Cox's estimates maximize the log-partial likelihood.

2. To analyze the effect of covariates, there is no need to estimate the nuisance parameter $h_0(t)$, the baseline hazard function.

3. Cox argues that most of the relevant information about the coefficients $\underline{\beta}$ for regression with censored data is contained in this partial likelihood.

4. This partial likelihood is not a true likelihood in that it does not integrate out to 1 over $\{0,1\}^n \times \Re_+^n$.

5. Censored individuals do not contribute to the numerator of each factor. But they do enter into the summation over the risk sets at death times that occur before a censored time.

6. Furthermore, this partial likelihood depends only on the ranking of the death times, since this determines the risk set at each death time. Consequently, inference about the effect of the explanatory variables on the hazard function depends only on the rank order of the death times! Here we see why this is often referred to as nonparametric. It only depends on the rank order! Look at the partial likelihood. There is no visible $t_{(j)}$ in the estimate for $\underline{\beta}$. It is a function of the $\underline{x}_{(j)}$'s which are determined by the rank order of the death times. So, the estimates are a function of the rank order of the death times.

We now present data diagnostic methods. We delay the examples and all S code until all relevant definitions and methods are presented.

6.3.1 Cox-Snell residuals for assessing the overall fit of a PH model

Recall from (1.6) the relationship

$$H(t) = -\log(S(t)) = -\log(1 - F(t)),$$

where F denotes the true d.f. of the survival time T and H denotes the true cumulative hazard rate. Also recall that regardless of the form of F, the random variable $F(T)$ is distributed uniformly on the unit interval $(0,1)$. Hence, the random variable $H(T)$ is distributed exponentially with hazard rate $\lambda = 1$. WHY! Let $\underline{x}_i$ denote the i-th individual's covariate vector. Then for a given $\underline{x}_i$, $H(t|\underline{x}_i)$ denotes the true cumulative hazard rate for an individual with covariate vector $\underline{x}_i$. It then follows

$$H(T_i|\underline{x}_i) \sim \exp(\lambda = 1).$$

Hence, if the Cox PH model is correct, then for a given $\underline{x}_i$, it follows

$$H(T_i|\underline{x}_i) = H_0(T_i) \times \exp\left(\sum_{j=1}^{m} \beta_j x_i^{(j)}\right) \sim \exp(\lambda = 1). \qquad (6.22)$$

The *Cox-Snell residuals* (Cox and Snell, 1968) are defined as

$$r_{Ci} = \widehat{H}_0(Y_i) \times \exp\left(\sum_{j=1}^{m} \widehat{\beta}_j x_i^{(j)}\right), i = 1, \ldots, n, \qquad (6.23)$$

where $Y_i = \min(T_i, C_i)$. The $\widehat{\beta}_j$'s are the *maximum partial likelihood estimates*, the estimates obtained from maximizing Cox's partial likelihood (6.21). The $\widehat{H}_0(t)$ is an empirical estimate of the cumulative hazard at time t. Typically this is either the Breslow or Nelson-Aalen estimate (page 33). S offers both with Nelson-Aalen as the default. For the definition of Breslow estimator, see Klein & Moeschberger (1997, page 237). If the final PH model is correct and the $\widehat{\beta}_j$'s are close to the true values of the β_j's, the r_{Ci}'s should resemble a censored sample from a unit exponential distribution. Let $H_E(t)$ denote the cumulative hazard rate of the unit exponential. Then $H_E(t) = t$. Let $\widehat{H}_{r_C}(t)$ denote a consistent estimator of the cumulative hazard rate of the r_{Ci}'s. Then $\widehat{H}_{r_C}(t)$ should be close to $H_E(t) = t$. Thus, for each uncensored r_{Ci}, $\widehat{H}_{r_C}(r_{Ci}) \approx r_{Ci}$. To check whether the r_{Ci}'s resemble a censored sample from a unit exponential, the plot of $\widehat{H}_{r_C}(r_{Ci})$ against r_{Ci} should be a 45°-line through the origin. See Figure 6.11.

Remarks:

1. The Cox-Snell residuals are most useful for examining the overall fit of a

model. A shortcoming is they do not indicate the type of departure from the model detected when the estimated cumulative hazard plot is not linear.

2. Ideally, the plot of $\widehat{H}_{r_C}(r_{C\,i})$ against $r_{C\,i}$ should include a confidence band so that significance can be addressed. Unfortunately, the $r_{C\,i}$ are not exactly a censored sample from a distribution. So this plot is generally used only as a rough diagnostic. A formal test of adequacy of the Cox PH model is given in Section 6.3.5.

3. The closeness of the distribution of the $r_{C\,i}$'s to the unit exponential depends heavily on the assumption that, when β and H_0 are replaced by their estimates, the probability integral transform $F(T)$ still yields uniform $(0,1)$ distributed variates. This approximation is somewhat suspect for small samples. Furthermore, departures from the unit exponential distribution may be partly due to the uncertainty in estimating the parameters β and H_0. This uncertainty is largest in the right-hand tail of the distribution and for small samples.

6.3.2 Martingale residuals for identifying the best functional form of a covariate

The martingale residual is a slight modification of the Cox-Snell residual. When the data are subject to right censoring and all covariates are time-independent (fixed at the start of the study), then the *martingale residuals*, denoted by $\widehat{M}_i$, are defined to be

$$\widehat{M}_i = \delta_i - \widehat{H}_0(Y_i) \times \exp\left(\sum_{j=1}^{m} \widehat{\beta}_j x_i^{(j)} \right) = \delta_i - r_{C\,i}, i = 1, \ldots, n, \qquad (6.24)$$

where $r_{C\,i}$ is the Cox-Snell residual.

These residuals are used to examine the best functional form for a given covariate using the assumed Cox PH model for the remaining covariates. Let the covariate vector $\underline{x}$ be partitioned into a $\underline{x}_*$ for which we know the functional form, and a single continuous covariate $x^{(1)}$ for which we are unsure of what functional form to use. We assume $x^{(1)}$ is independent of $\underline{x}_*$. Let $g(\cdot)$ denote the best function of $x^{(1)}$ to explain its effect on survival. The Cox PH model is then,

$$H(t|\underline{x}_*, x^{(1)}) = H_0(t) \times \exp\left(\underline{x}'_*\underline{\beta}_*\right) \times \exp\left(g(x^{(1)})\right), \qquad (6.25)$$

where $\underline{\beta}_*$ is an $m-1$ dimensional coefficient vector. To find $g(\cdot)$, we fit a Cox PH model to the data based on $\underline{x}_*$ and compute the martingale residuals, $\widehat{M}_i$, $i = 1, \ldots, n$. These residuals are plotted against the values $x_i^{(1)}$, $i = 1, \ldots, n$. A smoothed fit of the scatter plot is typically used. The smooth-fitted curve gives some indication of the function $g(\cdot)$. If the plot is linear, then no transformation of $x^{(1)}$ is needed. If there appears to be a threshold, then a discretized version of the covariate is indicated. The S function `coxph` provides martingale residuals as default and the S function `scatter.smooth` displays a smoothed fit of the scatter plot of the martingale residuals versus the covariate $x^{(1)}$. See Figure 6.12.

Remarks:

1. Cox-Snell residuals can be easily obtained from martingale residuals.

2. It is common practice in many medical studies to discretize continuous covariates. The martingale residuals are useful for determining possible cut points for such variables. In Subsection 6.3.8 we present a cut point analysis with bootstrap validation conducted for the variable KPS.PRE. in the CNS data.

3. The martingale residual for a subject is the difference between the observed and the expected number of deaths for the individual. This is so because we assume that no subjects can have more than one death and the second factor in expression (6.24) is the estimated cumulative hazard of death for the individual over the interval $(0, y_i)$.

4. The martingale residuals sum to zero; that is, $\sum_{i=1}^{n} \widehat{M}_i = 0$. For "large" n, the $\widehat{M}_i$'s are an uncorrelated sample from a population with mean zero. However, they are not symmetric around zero because the martingale residuals take values between $-\infty$ and 1.

5. For the more general definition of the martingale residuals which includes time-dependent covariates, see Klein & Moeschberger (1997, pages 333 and 334). On page 337 under *Theoretical Notes* these authors further explain why a smoothed plot of the martingale residuals versus a covariate should reveal the correct functional form for including $x^{(1)}$ in a Cox PH model.

6.3.3 Deviance residuals to detect possible outliers

These residuals were defined and discussed in great detail in the previous section on diagnostic methods for parametric models. Except for a slight modification in the definition of deviance, all plots and interpretations carry over. What's different here is that we no longer have a likelihood. We are working with a partial likelihood. However, we may still define deviance analogously, using the partial likelihood. All tests and their large sample distributions still apply. The deviance residual is used to obtain a residual that is more symmetrically shaped than a martingale residual as the martingale residual can be highly skewed. The *deviance residual* (Therneau, Grambsch, and Fleming, 1990) is defined by

$$D_i = \text{sign}(\widehat{M}_i) \times \sqrt{-2 \times \left(\widehat{M}_i + \delta_i \log(\delta_i - \widehat{M}_i) \right)}, \qquad (6.26)$$

where $\widehat{M}_i$ is the martingale residual defined in Subsection 6.3.2. The log function inflates martingale residuals close to one, while the square root contracts the large negative martingale residuals. In all plots, potential outliers correspond to large absolute valued deviance residuals. See Figure 6.13.

Remarks:

1. Therneau, Grambsch, and Fleming (1990) note "When censoring is minimal, less than 25% or so, these residuals are symmetric around zero. For censoring greater than 40%, a large bolus of points with residuals near zero distorts the normal approximation but the transform is still helpful in symmetrizing the set of residuals." Obviously, deviance residuals do not necessarily sum to zero.

2. Type `resid(fit,type="deviance")`, where `fit` is the `coxph` object, to obtain these residuals.

6.3.4 Schoenfeld residuals to examine fit and detect outlying covariate values

The kth *Schoenfeld residual* (Schoenfeld, 1982) defined for the kth subject on the jth explanatory variable $x^{(j)}$ is given by

$$r_{s_{jk}} = \delta_k \{ x_k^{(j)} - a_k^{(j)} \}, \tag{6.27}$$

where δ_k is the kth subject's censoring indicator, $x_k^{(j)}$ is the value of the jth explanatory variable on the kth individual in the study,

$$a_k^{(j)} = \frac{\sum_{m \in \mathcal{R}(y_k)} \exp(\underline{x}'_m \widehat{\underline{\beta}}) x_m^{(j)}}{\sum_{m \in \mathcal{R}(y_k)} \exp(\underline{x}'_m \widehat{\underline{\beta}})},$$

and $\mathcal{R}(y_k)$ is the risk set at time y_k. The MLE $\widehat{\underline{\beta}}$ is obtained from maximizing the Cox's partial likelihood function $L_c(\underline{\beta})$ (6.21). Note that nonzero residuals only arise from uncensored observations.

We see this residual is just the difference between $x_k^{(j)}$ and a weighted average of the values of explanatory variables over individuals at risk at time y_k. The **weight** used for the mth individual in the risk set at y_k is

$$\frac{\exp(\underline{x}'_m \widehat{\underline{\beta}})}{\sum_{m \in \mathcal{R}(y_k)} \exp(\underline{x}'_m \widehat{\underline{\beta}})},$$

which is the contribution from this individual to the maximized partial likelihood (6.21). Further, since the MLE of $\underline{\beta}$, $\widehat{\underline{\beta}}$, is such that

$$\frac{\partial \log \left(L_c(\underline{\beta}) \right)}{\partial \beta_j} \bigg|_{\widehat{\underline{\beta}}} = 0,$$

the Schoenfeld residuals for each predictor $x^{(j)}$ must sum to zero. These residuals also have the property that in large samples the expected value of $r_{s_{jk}}$ is zero and they are uncorrelated with each other. Furthermore, suppose y_k is a small failure time relative to the others. Then its risk set is huge. Hence, in general not only do subjects in the risk set have a wide range of covariate values, but also the weight assigned to each covariate value associated with the risk set is small. Therefore, individuals with large covariate values who die at early failure times would have large positive Schoenfeld residuals. This can be most easily seen if we rewrite $r_{s_{jk}}$ (6.27) as

$$
x_k^{(j)} \left(1 - \frac{\exp(\underline{x}_k' \widehat{\underline{\beta}})}{\sum_{m \in \mathcal{R}(y_k)} \exp(\underline{x}_m' \widehat{\underline{\beta}})} \right) - \sum_{l \in \mathcal{R}(y_k); \, l \neq k} \left(x_l^{(j)} \frac{\exp(\underline{x}_l' \widehat{\underline{\beta}})}{\sum_{m \in \mathcal{R}(y_k)} \exp(\underline{x}_m' \widehat{\underline{\beta}})} \right).
$$

$$(6.28)$$

It is clear from expression (6.28) that the first term is large and the second term is small relative to the first term. Similarly, the individuals with small covariate values who die at early failure times would have large negative Schoenfeld residuals. WHY! Therefore, a few relatively large absolute valued residuals at early failure times may not cause specific concern. Thus, these residuals are helpful in detecting outlying covariate values for early failure times. However, if the PH assumption is satisfied, large Schoenfeld residuals are not expected to appear at late failure times. WHY! Therefore, we should check the residuals at late failure times. See Figure 6.14.

Remarks:

1. Schoenfeld calls these residuals the partial residuals as these residuals are obtained from maximizing the partial likelihood function. Collett (1994, page 155), among others, calls these residuals the score residuals as the first derivative of the partial likelihood can be considered as the efficient score.

2. Use `coxph.detail` to obtain the detailed `coxph` object. This includes ranked observed times along with a corresponding censoring status vector and covariate information.

3. Type `resid(fit,type="schoenfeld")`, where `fit` is the `coxph` object, to obtain these residuals. `coxph` does not output the value of Schoenfeld residual for subjects whose observed survival time is censored as these are zeros.

4. If the assumption of proportional hazards holds, a plot of these residuals against ordered death times should look like a tied down random walk. Otherwise, the plot will show too large residuals at some times.

6.3.5 Grambsch and Therneau's test for PH assumption

As an alternative to proportional hazards, Grambsch and Therneau (1994) consider time-varying coefficients $\beta(t) = \beta + \underline{\theta}g(t)$, where $g(t)$ is a predictable process (a postulated smooth function). Given $g(t)$, they develop a score test for $H_0 : \underline{\theta} = \underline{0}$ based on a generalized least squares estimator of $\underline{\theta}$. Defining *scaled Schoenfeld residuals* by the product of the inverse of the estimated variance-covariance matrix of the kth Schoenfeld residual and the kth Schoenfeld residual, they show the kth scaled Schoenfeld residual has approximately mean $\underline{\theta}g(t_k)$ and the kth Schoenfeld residual has an easily computable variance-covariance matrix. Motivated by these results, they also develop a graphical method. They show by Monte Carlo simulation studies that a smoothed scatter plot of $\widehat{\beta}(t_k)$, the kth scaled Schoenfeld residual plus $\widehat{\beta}$ (the maximum partial likelihood estimate of β), versus t_k reveals the functional form of $\beta(t)$. Under H_0, we expect to see a constant function over time. Both of these can be easily done with the S functions cox.zph and plot. See Figure 6.15.

Remarks:

1. The function $g(t)$ has to be specified. The default in the S function cox.zph is K-M(t). The options are $g(t) = t$ and $g(t) = \log(t)$ as well as a function of one's own choice.

2. plot(out), where out is the cox.zph object, gives a plot for each covariate. Each plot is of a component of $\widehat{\beta}(t)$ versus t together with a spline smooth and ± 2 s.e. pointwise confidence bands for the spline smooth.

3. A couple of useful plots for detecting violations of the PH assumption are recommended:

   (a) A plot of log-cumulative hazard rates against time is useful when x is a group variable. For example, if there are two treatment groups, plot both curves on the same graph and compare them. If the curves are parallel over time, it supports the PH assumption. If they cross, this is a blatant violation.

   (b) A plot of differences in log-cumulative hazard rates against time is also useful. This plot displays the differences between the two curves in the previous graph. If the PH assumption is met, this plot is roughly constant over time. Otherwise, the violation will be glaring. This plot follows Miller's basic principle discussed here on page 144.

6.3.6 dfbetas to assess influence of each observation

Here we want to check the influence of each observation on the estimate $\widehat{\beta}$ of the $\underline{\beta}$. Let $\widehat{\underline{\beta}}_{(k)}$ denote the estimated vector of coefficients computed on the sample with the kth observation deleted. Then we check which components of the vector $\widehat{\beta} - \widehat{\beta}_{(k)}$ have unduly large absolute values. Do this for each of the n observations. One might find this measure similar to dfbetas in the linear regression. This involves fitting $n+1$ Cox regression models. Obviously, this is computationally expensive unless the sample size is small. Fortunately, there exists an approximation based on the Cox PH model fit obtained from the whole data that can be used to circumvent this computational expense. The kth *dfbeta* is defined as

$$\text{dfbeta}_k = I(\widehat{\beta})^{-1}(r^*_{s_{1k}}, \ldots, r^*_{s_{mk}})', \tag{6.29}$$

where $I(\widehat{\beta})^{-1}$ is the inverse of the observed Fisher information matrix, and for $j = 1, \ldots, m$,

$$r^*_{s_{jk}} = \delta_k\{x_k^{(j)} - a_k^{(j)}\} - \exp(\underline{x}'_k\widehat{\beta}) \sum_{t_i \leq y_k} \frac{\{x_k^{(j)} - a_i^{(j)}\}}{\sum_{l \in \mathcal{R}(t_i)} \exp(\underline{x}'_l\widehat{\beta})}.$$

Note that the first component is the kth Schoenfeld residual and the second component measures the combined effect over all the risk sets that include the kth subject. This expression, proposed by Cain and Lange (1984), well approximates the difference $\widehat{\beta} - \widehat{\beta}_{(k)}$ for $k = 1, \ldots, n$. The authors note that the above two components in general have opposite signs. The second component increases in absolute magnitude with t_k, as it is the sum of an increasing number of terms. Thus, for early death times, the first component dominates, while for later death times, the second is usually of greater magnitude. This means that for patients who die late, the fact that the patient lived a long time, and thus was included in many risk sets, has more effect upon $\widehat{\beta}$ than does the fact that the patient died rather than was censored. Plots of these quantities against the case number (index) or against their respective covariate $x_k^{(j)}$ are used to gauge the influence of the kth observation on the jth coefficient. See Figure 6.16.

Remarks:

1. The S function `resid(fit,type="dfbetas")` computes dfbeta divided by the s.e.'s for the components of $\widehat{\beta}$, where `fit` is the `coxph` object.

2. Collett (1994) calls these standardized delta-beta's.

3. There are a number of alternate expressions to expression (6.29). For example, see pages 359 through 365 in Klein & Moeschberger (1997).

4. This measure is analogous to the measures of influence for ordinary linear regression developed by Belsley *et al.* (1980) and Cook and Weisberg (1982).

6.3.7 CNS lymphoma example: checking the adequacy of the PH model

We apply some model checking techniques on the final reduced model
cns2.coxint6.

```
# Cox-Snell residuals for overall fit of a model are not
# provided directly by coxph object. You can derive them
# from the martingale residuals which are the default
# residuals.
```

Figure 6.11:

```
> attach(cns2)
> rc <- abs(STATUS - cns2.coxint6$residuals) # Cox-Snell
                                             # residuals!
> km.rc <- survfit(Surv(rc,STATUS) ~ 1)
> summary.km.rc <- summary(km.rc)
> rcu <- summary.km.rc$time # Cox-Snell residuals of
                            # uncensored points.
> surv.rc <- summary.km.rc$surv
> plot(rcu,-log(surv.rc),type="p",pch=".",
xlab="Cox-Snell residual rc",ylab="Cumulative hazard on rc")
> abline(a=0,b=1); abline(v=0); abline(h=0)

  # The martingale residual plots to check functional form of
  # covariate follow.
```

Figure 6.12:

```
> fit <- coxph(Surv(B3TODEATH,STATUS) ~ GROUP+SEX+AGE60+
               SEX:AGE60)
> scatter.smooth(cns2$KPS.PRE.,resid(fit),type="p",pch=".",
    xlab="KPS.PRE.",ylab="Martingale residual")

  # The deviance residual plots to detect outliers follow:
```

Figure 6.13:

```
> dresid <- resid(cns2.coxint6,type="deviance") # deviance
                                                # residual
> plot(dresid,type="p",pch=".")
> abline(h=0)
> plot(B3TODEATH,dresid,type="p",pch=".")
> abline(h=0)
> plot(GROUP,dresid,type="p",pch=".")
> abline(h=0)
> plot(SEX,dresid,type="p",pch=".")
```

```
> abline(h=0)
> plot(AGE60,dresid,type="p",pch=".")
> abline(h=0)
> plot(KPS.PRE.,dresid,type="p",pch=".")
> abline(h=0)

# Schoenfeld residuals to examine fit and detect outlying
# covariate values
```

Figure 6.14:

```
> detail <- coxph.detail(cns2.coxint6) # detailed coxph object
> time <- detail$y[,2] # ordered times including censored ones
> status <- detail$y[,3] # censoring status
> sch <- resid(cns2.coxint6,type="schoenfeld") # Schoenfeld
                                                # residuals
> plot(time[status==1],sch[,1],xlab="Ordered survival time",
   ylab="Schoenfeld residual for KPS.PRE.") # time[status==1]
          # is the ordered uncensored times and sch[,1] is the
          # resid for KPS.PRE.

# The scaled Schoenfeld residuals and the Grambsch and
# Therneau's test for time-varying coefficients to assess
# PH assumption follow:
```

Figure 6.15:

```
> PH.test <- cox.zph(cns2.coxint6)
> PH.test
             rho   chisq     p
KPS.PRE.  0.0301  0.025  0.874
   GROUP  0.1662  1.080  0.299
     SEX  0.0608  0.103  0.748
   AGE60 -0.0548  0.114  0.736
SEX:AGE60 0.0872  0.260  0.610
  GLOBAL      NA  2.942  0.709

> par(mfrow=c(3,2)); plot(PH.test)

# The dfbetas is approximately the change in the
# coefficients scaled by their standard error. This
# assists in detecting influential observations on
# the estimated beta coefficients.
```

Figure 6.16:

```
> par(mfrow=c(3,2))
> bresid <- resid(cns2.coxint6,type="dfbetas")
> index <- seq(1:58)
> plot(index,bresid[,1],type="h",ylab="scaled change in coef",
        xlab="observation")
> plot(index,bresid[,2],type="h",ylab="scaled change in coef",
        xlab="observation")
> plot(index,bresid[,3],type="h",ylab="scaled change in coef",
        xlab="observation")
> plot(index,bresid[,4],type="h",ylab="scaled change in coef",
        xlab="observation")
> plot(index,bresid[,5],type="h",ylab="scaled change in coef",
        xlab="observation")

# For the sake of comparison, we consider the scaled
# Schoenfeld residuals and the test for time-varying
# coefficients for the main effects model cns2.cox3.
```

Figure 6.17:

```
> PHmain.test <- cox.zph(cns2.cox3)
> PHmain.test
             rho   chisq      p
KPS.PRE.  0.0479  0.0671  0.796
   GROUP  0.1694  1.1484  0.284
     SEX  0.2390  1.9500  0.163
  GLOBAL      NA  3.1882  0.364

> par(mfrow=c(2,2)); plot(PHmain.test)
> detach()
```

Results:

- We see from the Cox-Snell residual plot, Figure 6.11, that the final model gives a reasonable fit to the data. Overall the residuals fall on a straight line with an intercept zero and a slope one. Further, there are no large departures from the straight line and no large variation at the right-hand tail.

- In the plot of the Martingale residuals, Figure 6.12, there appears to be a bump for KPS.PRE. between 80 and 90. However, the lines before and after the bump nearly coincide. Therefore, a linear form seems appropriate for KPS.PRE. There are occasions where a discretized, perhaps dichotomized, version of a continuous variable is more appropriate and informative. See an extensive cut point analysis conducted in the next subsection.

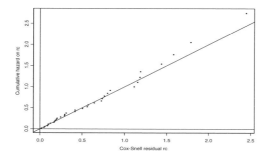

Figure 6.11 *Cox-Snell residuals to assess overall model fit.*

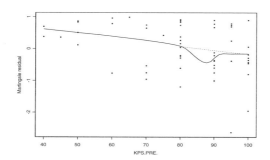

Figure 6.12 *Martingale residuals to look for best functional form of the continuous covariate* KPS.PRE.

- The deviance residual plot, Figure 6.13, shows a slight tendency for larger survival times to have negative residuals. This suggests that the model overestimates the chance of dying at large times. However, there is only one possible outlier at the earliest time and this may not cause concern about the adequacy of the model. All the other plots in the same figure show that the residuals are symmetric around zero and there is at most one possible outlier.

- The subjects with the largest absolute valued Schoenfeld residuals for KPS.PRE. are 40, 8, 35, and 11. These subjects have very early failure times .125, .604, .979, and 1.375 years and are the patients who have either the largest or the smallest KPS.PRE. values. Thus, these residuals do not cause specific concern. The plots for the other covariates are not shown here. But all of them show no large residuals. Therefore, the PH assumption seems to be appropriate.

- The results from the test for constancy of the coefficients based on scaled Schoenfeld residuals indicate the PH assumption is satisfied by all five co-

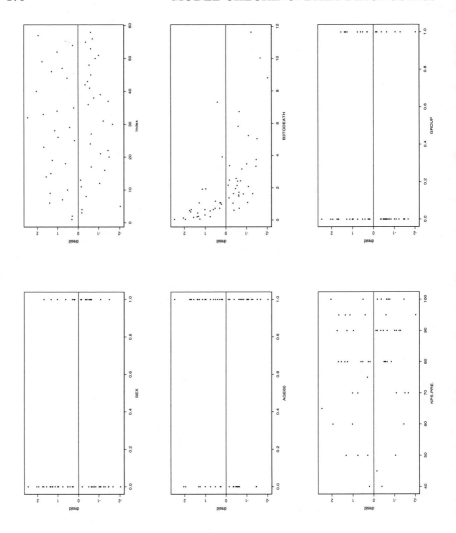

Figure 6.13 *Deviance residuals to check for outliers.*

variates in the model with all p-values being at least 0.299. Figure 6.15 also supports that the PH assumption is satisfied for all the covariates in the model.

- For the sake of comparison, we consider the main effects model, cns2.cox3, as well. Although the results from the test for constancy of the coefficients indicate that the PH assumption is satisfied by all three covariates in the model with all p-values being at least 0.16, Figure 6.17 gives some mild evidence that the PH assumption may be violated for the GROUP and SEX variables. This results from the fact that in this model there are no

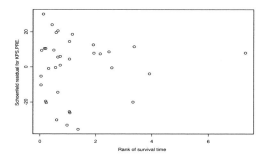

Figure 6.14 *Schoenfeld residuals for* KPS.PRE. *against ordered survival times.*

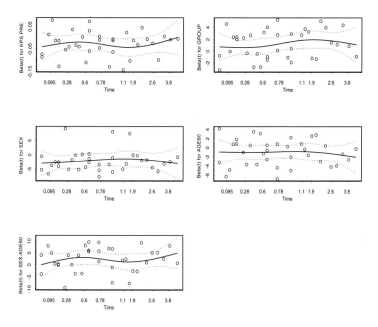

Figure 6.15 *Diagnostic plots of the constancy of the coefficients in cns2.coxint6. Each plot is of a component of $\widehat{\beta}(t)$ against ordered time. A spline smoother is shown, together with ± 2 standard deviation bands.*

interaction effect terms when there is a significant interaction effect between SEX and AGE60 as evidenced by the model cns2.coxint6. This again tells us how important it is to consider interaction effects in modelling.

• The plot of the dfbetas, Figure 6.16, shows that most of the changes in the

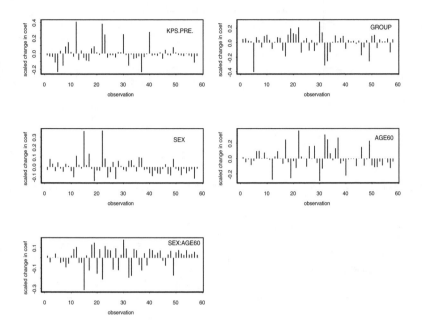

Figure 6.16 *The dfbetas to detect influential observations on the five estimated coefficients corresponding to the predictors.*

regression coefficients are less than .3 s.e.'s of the coefficients and all others are less than .4 s.e.'s. Therefore, we conclude that there are no influential subjects.

6.3.8 Cut point analysis with bootstrap validation

We perform cut point analysis by dichotomizing the continuous covariate KPS.PRE. We define the indicator variable K by

$$K = \begin{cases} 0 & \text{if } \text{KPS.PRE.} < \theta \\ 1 & \text{if } \text{KPS.PRE.} \geq \theta. \end{cases}$$

Before we proceed, we introduce the **profile likelihood function**. Suppose that the parameter in a model may be partitioned as (θ, η) where θ is the parameter of interest and η is the nuisance parameter. Given a likelihood $L(\theta, \eta | x_1, \ldots, x_n)$, the profile likelihood for θ is defined as the function

$$\theta \rightarrow \sup_{\eta} L(\theta, \eta | x_1, \ldots, x_n), \qquad (6.30)$$

where the supremum is taken over all possible values of η and $x_1, \ldots, x_n$ are the realizations of the iid random variables $X_1, \ldots, X_n$. This definition

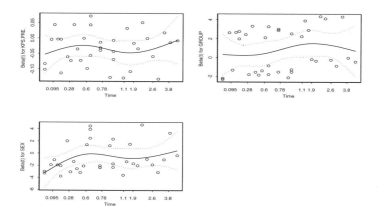

Figure 6.17 *Diagnostic plots of the constancy of the coefficients in cns2.cox3. Each plot is of a component of $\widehat{\beta}(t)$ against ordered time. A spline smoother is shown, together with ± 2 standard deviation bands.*

is quoted from van der Vaart (1998) where he also illustrates examples in semiparametric survival models.

In the following, we treat the values of KPS.PRE. as the values of θ and the β in the Cox PH model as η in the above usual setting. For various values of θ we fit a Cox PH model with the five covariates GROUP, SEX, AGE60, SEX:AGE60, and K. For each θ value, a profile log-likelihood value is obtained. The value of θ which maximizes the profile likelihood function is the desired cut point (threshold). The maximum must occur at one of the 58 KPS.PRE. scores.

The function `cutpt.coxph` computes profile log-likelihoods for various quantiles. Before using this function, put the variable to be dichotomized in the first column of the data frame (use `move.col`). Then generate an additional column of Bernoulli values as many as the number of rows. This creates a place holder for the variable K. Run `coxph` on the model specified above. The output from `coxph` provides the necessary formula within the function `cutpt.coxph` automatically. `cutpt.coxph` has three arguments: (object,data,q), where `object` is a `coxph` object, `data` is a data frame, and `q` is a vector of various quantiles. If `q` is not provided, the default quantiles, seq(.1,.9,.1), are used. We see from Figure 6.18 that the best choice of θ is 90.

```
> move.col(cns2,1,cns2,9) # Moves KPS.PRE. in col. 9 to
                          # col. 1.
> attach(cns2)
> temp <- coxph(Surv(B3TODEATH,STATUS) ~ GROUP+SEX+AGE60+
                SEX:AGE60+K) # provides the necessary
```

```
                # formula within cutpt.coxph automatically.
> cns2cutpt <- cutpt.coxph(object=temp,data=cns2,
                           q=seq(0.05,0.95,0.05))
                           # output of cutpt.coxph
> plot(cns2cutpt$out,type="l",lwd=2,col=1,xlab="KPS.PRE.",
    ylab="Profile log-likelihood") # plot of quantiles
        # versus corresponding profile likelihoods
```

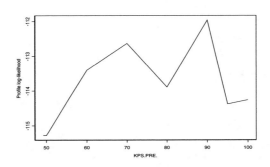

Figure 6.18 *Profile log-likelihoods at quantiles of* KPS.PRE.

Now we fit the Cox PH model with GROUP, SEX, AGE60, SEX:AGE60, and K where

$$K = \begin{cases} 0 & \text{if KPS.PRE.} < 90 \\ 1 & \text{if KPS.PRE.} \geq 90. \end{cases}$$

A plot of the two estimated survival curves for $K = 1$ and 0, each evaluated at GROUP $= 0$, SEX $= 0$ (male), AGE60 $= 1$, and SEX:AGE60 $= 0$, is presented in Figure 6.19.

```
> cns2$K <- as.integer(cns2$KPS.PRE.>=90) # creates the
                                # indicator variable K
> cns2cutpt1 <- coxph(Surv(B3TODEATH,STATUS) ~ GROUP+SEX+
                      AGE60+SEX:AGE60+K)
> summary(cns2cutpt1)
```

```
  n= 58
              coef exp(coef) se(coef)      z       p
    GROUP    1.285     3.616    0.373   3.45 0.00056
      SEX   -1.602     0.202    0.676  -2.37 0.01800
    AGE60   -0.713     0.490    0.481  -1.48 0.14000
        K   -1.162     0.313    0.420  -2.77 0.00560
SEX:AGE60    1.582     4.864    0.864   1.83 0.06700

Rsquare=0.374   (max possible=0.987)
Likelihood ratio test = 27.2 on 5 df, p=0.0000524
```

```
Wald test = 23.6 on 5 df, p=0.000261
Score (logrank) test = 27 on 5 df, p=0.0000576
> surv.cutpt11 <- survfit(cns2cutpt1,data.frame(GROUP=0,
                         SEX=0,AGE60=1,K=0))
> surv.cutpt12 <- survfit(cns2cutpt1,data.frame(GROUP=0,
                         SEX=0,AGE60=1,K=1))
> plot(surv.cutpt11,type="l",lwd=2,col=1,lty=1,
       conf.int=F,lab=c(10,7,7),yscale=100,
       xlab="Survival Time in Years from First BBBD",
       ylab="Percent Alive")
> lines(surv.cutpt12$time,surv.cutpt12$surv,type="s",
        lty=4,col=1,lwd=2)
> detach()
```

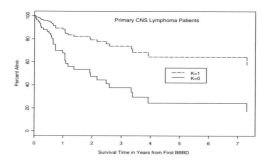

Figure 6.19 *Estimated Cox survival curves for K with cut point = 90.*

We summarize the output from the final model and the dichotomized model in Table 6.1.

Table 6.1: *Summary of fits with continuous versus dichotomized KPS.PRE.*

	Continuous KPS.PRE.			Dichotomized KPS.PRE.=K	
	coefficient	p-value		coefficient	p-value
GROUP	1.1592	0.0022	GROUP	1.285	0.00056
SEX	−2.1113	0.0026	SEX	−1.602	0.01800
AGE60	−1.0538	0.0210	AGE60	−0.713	0.14000
SEX:AGE60	2.1400	0.0120	SEX:AGE60	1.582	0.06700
KPS.PRE.	−0.0307	0.0028	K	−1.162	0.00560
LRT = 27.6		.00004	LRT = 27.2		0.00005

Results:

- According to the LRT criterion in Table 6.1, the overall fit is not improved by using K.

- The variable K improves the effect of GROUP variable, whereas it makes the effects of all others weaker. In particular, AGE60 and SEX:AGE60 have p-values .140 and .067. This is contradictory to the clear interaction effect shown in Figures 5.2 and 5.3. Therefore, dichotomizing KPS.PRE. is not recommended.

Bootstrap validation to check robustness

The bootstrap method was introduced by Efron (1979). For censored survival data, see Davison and Hinkley (1999). For general introduction, see either Lunneborg (2000) or Efron and Tibshirani (1993). The former is appropriate for scientific researchers without a mathematical background, while the latter is appropriate for researchers with a mathematical background. Here we apply so called nonparametric bootstrap sampling. Suppose we have n subjects in the study. Then select n subjects from the original data with replacement with equal probability. The data values corresponding to these selected individuals make a bootstrap sample. Run your model with this bootstrap sample and keep the statistics of interest. Repeat this process B times where $B = 1000$ is used most often but the larger the better in general.

Often, as the case in the CNS study, the continuous variable under cut point analysis has many repeated values. Furthermore, when at the extreme (minimum or maximum) values this sets either all ones or zeros. This causes many singularities. A remedy for this is not to use all 58 values but to use the sample quantiles of each bootstrap sample. In this study we use the .05, .10, ..., .95 quantiles. The bootstrap density histogram, Figure 6.20, shows a very similar shape to the profile likelihood in Figure 6.18. This demonstrates the cut point analysis is robust.

We use the S-PLUS function `bootstrap`. This function is not available in R. Venables and Ripley (2002, page 464) provides useful S library sources including boot library. One can download boot library prepackaged for Windows users (boot.zip) from "http://www.stats.ox.ac.uk/pub/MASS4/Winlibs". The equivalent R packages are available from http://lib.stat.cmu.edu/R/CRAN/

```
> temp1 <- bootstrap(cns2,cutpt.coxph(object=cns2cutpt1,
            data=cns2,q=seq(.05,.95,.05))$cutpt,B=1000)
            # Bootstrap output
> plot(temp1,xlab="cut points",ylab="density",
    main="Bootstrap Density Histogram of Cut Points")
    # Density histogram of bootstrap profile
    # log-likelihoods at 19 quantiles
```

There are two peaks on this graph. So we tried 70 and 90 as possible cut points in the final model, cns2.coxint6. But the fit was even worse than that of the model with dichotomized KPS.PRE. This suggests that there has to be a dominant peak in the profile likelihood.

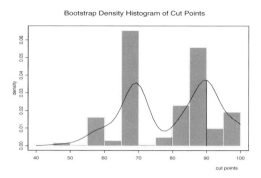

Figure 6.20 *Bootstrap density histogram of the maximum profile log-likelihood estimates of θ.*

For the sake of comparison, we consider the martingale residual plot and cut point analysis based on the main effects model, cns2.cox3. Here we only report the results because the commands needed are similar to those used for the interaction model. The bootstrap density histogram, Figure 6.23, shows a very

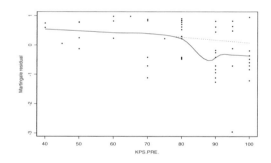

Figure 6.21 *Martingale residuals to look for best functional form of the continuous covariate KPS.PRE. in main effects model cns2.cox3.*

Table 6.2: *Summary of fits with continuous versus dichotomized KPS.PRE.*

Continuous KPS.PRE.			Dichotomized KPS.PRE.$=K$		
	coefficient	p-value		coefficient	p-value
GROUP	.7785	.028	GROUP	1.089	.00031
SEX	$-.7968$	.052	SEX	$-.665$	.086
KPS.PRE.	$-.0347$	.0056	K	-1.396	.00024
LRT $= 20.3$		.000146	LRT $= 23.5$		.000032

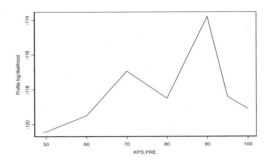

Figure 6.22 *Profile log-likelihoods at quantiles of* KPS.PRE. *in main effects model* cns2.cox3.

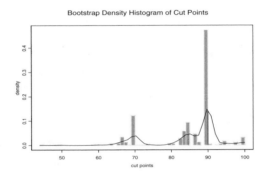

Figure 6.23 *Bootstrap density histogram of the maximum profile likelihood estimates of θ in main effects model* cns2.cox3.

similar shape to the profile log-likelihood in Figure 6.22. This again demonstrates the robustness of the cut point analysis. More importantly, it shows a dominant peak, which suggests that the cut point should be significant. That is, the cut point must improve the model.

Results:

- The martingale residual plot in Figure 6.21 shows a clear drop after about KPS.PRE. = 80 and the drop is maintained unlike that in Figure 6.12.

- According to the LRT criterion in Table 6.2, the overall fit is improved using the variable K, the dichotomized version of KPS.PRE. This supports our conjecture that a dominant peak in the profile log-likelihood guarantees the existence of a significant cut point.

- GROUP becomes stronger, whereas SEX becomes a bit weaker. SEX was originally marginally significant (p-value $= .052$) in the main effects model and so this does not cause specific concern.

- The variable K has a very small p-value, .00024. Furthermore, a more pronounced effect on survival is noticed although the plot is not shown here.

Discussion

- The martingale residual plot in Figure 6.21 based on the main effects model cns2.cox3 shows two parallel lines, whereas the plot in Figure 6.12 for the two-way interaction model cns2.coxint6 shows only a single line with a little bump. Unlike the bump, the parallel lines indicate that the cut point exists between the end of one line and the start of the other line. From this CNS data, we find that as long as there are parallel lines, the vertical distance between the lines doesn't matter. Furthermore, if there is no room for improvement, there will be a little bump on the martingale residual plot. Therefore, the martingale residual plot is highly informative in finding possible cut points.

- Cut point analysis may be useful when we are still trying to improve our current model. This is illustrated in the CNS example. In the main effects model there is room for improvement. By adding the dichotomized variable K, the new model improves not only the strength of the variable GROUP but the overall model fit. The interaction model only displays a less drastic bump since it has no room for improvement.

- Bootstrap validation demonstrates the robustness of the cut point analysis. See Figures 6.20 and 6.23.

6.4 Exercises

A. *Applications*

6.1 Refer to Exercise 5.1(a). Using the same model, generate versions of Figures 6.11 to 6.16. Comment on each of your plots.

6.2 (Continued from Exercise 6.1) Do the cut point analysis for the covariate alter. Compare the results of this analysis with those in Exercise 5.1(a).

Hint: See the code on pages 173 to 176.

B. *Theory and WHY!*

6.3 Verify expression (6.10).

6.4 Verify expression (6.14).

6.5 Verify expression (6.16).

6.6 Prove expression (6.18).

6.7 Answer the WHY! on page 152.

6.8 Prove the WHY! on page 159.

6.9 Answer the WHY!'s on page 163.

## CHAPTER 7

Additional Topics

7.1 Extended Cox model

This section describes how the Cox PH model can be extended to allow time-dependent variables as predictors. The bulk of the material presented here is an abridged version of Kleinbaum (1996, pages 216–235). In addition, we lightly touch on the usefulness of the counting process formulation of the PH model. We describe how to reformulate, and reenter, a data set to match the counting process formulation needed to implement an extended Cox model properly using the S function `coxph`.

Objectives of this section:

After studying Section 7.1, the student should:

1. Know how to construct extended Cox model form.

2. Know characteristics of this model.

3. Know the formula and interpretation of the hazard ratio (HR).

4. Be able to identify several types of time-dependent predictor variables.

5. Know how to implement the extended Cox model using S.

Three types of time-dependent predictor variables:

1. "defined"

2. "internal"

3. "ancillary" (also referred to as "external")

A **time-dependent** variable is defined as any variable whose value for a given subject may vary over time t. This is in contrast to a **time-independent** variable (assumed in the PH model) whose value remains constant over time. One simple example of a time-independent variable is RACE. Two examples of **"defined" time-dependent** variables follow. Most "defined" variables are of the form of the product of a time-independent variable multiplied by time or some function of time $g(t)$.

1. $E \times (\log t - 3)$, where E denotes, say, a $(0,1)$ exposure status variable determined at one's entry into the study and $g(t) = \log(t) - 3$.

2. $E \times g(t)$ where

$$g(t) = \left\{ \begin{array}{ll} 1 & \text{if } t \geq t_0 \\ 0 & \text{if } t < t_0. \end{array} \right.$$

The function $g(t)$ is called a "heavyside" function. We illustrate later how these types of functions may be used as one method for the analysis when a time-independent variable like E does not satisfy the PH assumption.

Another type of time-dependent variable is called an **"internal"** variable. The values change over time for any subject under study. The reason for a change depends on "internal" characteristics or behavior specific to the individual. Some examples are:

1. exposure level E at time t; $E(t)$,

2. employment status (EMP) at time t; $EMP(t)$,

3. smoking status (SMK) at time t; $SMK(t)$, and

4. obesity level (OBS) at time t; $OBS(t)$.

In contrast, a variable is called an **"ancillary"** variable if its values change primarily because of **"external"** characteristics of the environment that may affect several individuals simultaneously. Two such examples of this type are:

1. air pollution index at time t for a particular geographical area, and

2. EMP at time t, if the primary reason for the status depends more on general economic circumstances than on individual characteristics.

We give one more example which may be part internal and part external. Consider "heart transplant status" HT at time t for a person identified to have a serious heart condition, making him or her eligible for a transplant. The variable HT is defined as:

$$HT(t) = \left\{ \begin{array}{ll} 1 & \text{if person already received a transplant at some time } t_0 \leq t \\ 0 & \text{if person did not receive a transplant by time } t; \text{ i.e., } t_0 > t. \end{array} \right.$$

Note that once a person receives a transplant, at time t_0, the value of HT remains at 1 for all time points thereafter. In contrast, a person who never receives a transplant has HT equal to 0 for all times during the period he or she is in the study. $HT(t)$ can be considered essentially an internal variable because individual traits of an eligible transplant recipient are important determinants of the decision to carry out transplant surgery. Nevertheless, availability of a donor heart prior to tissue and other matching with an eligible recipient can be considered as an "ancillary" characteristic external to the recipient.

Note: Computer commands differ for defined versus internal versus ancillary.

But, the form of the extended Cox model and the procedures for analysis are the same regardless of the variable type.

The extended Cox model:

We write the extended Cox model that incorporates both time-independent and time-dependent variables as follows:

$$h(t|\underline{x}(t)) = h_0(t) \exp\left(\sum_{i=1}^{p_1} \beta_i x^{(i)} + \sum_{j=1}^{p_2} \gamma_j x^{(j)}(t)\right)$$

$$\underline{x}(t) = \left(\underbrace{x^{(1)}, x^{(2)}, \cdots, x^{(p_1)}}_{\text{time}-\text{independent}}, \underbrace{x^{(1)}(t), x^{(2)}(t), \cdots, x^{(p_2)}(t)}_{\text{time}-\text{dependent}}\right),$$

where β_i and γ_j are regression coefficients corresponding to covariates, $h_0(t)$ is a baseline hazard function, and $\underline{x}(t)$ denotes the entire collection of covariates at time t. A simple example with one time-independent variable and one time-dependent variable is given by

$$h(t|\underline{x}(t)) = h_0(t) \exp\left(\beta E + \gamma(E \times t)\right),$$

$$p_1 = 1, \quad p_2 = 1,$$

$$\underline{x}(t) = \left(x^{(1)} = E, \quad x^{(1)}(t) = E \times t\right).$$

The estimates are obtained by maximizing a partial likelihood function L. The computations are more complicated than for the Cox PH model, because the risk sets used to form L are more complicated with time-dependent variables.

Methods for making statistical inference are essentially the same as for the PH model. One can use Wald and/or LRT's and large sample confidence interval methods.

An important assumption of the extended Cox model is that the effect of a time-dependent variable $X^{(j)}(t)$ on the survival probability at time t depends on the value of this variable at the *same* time t, and not on the value at an earlier or later time. However, it is possible to modify the definition of the time-dependent variable to allow for a **"lag-time" effect.** For example, suppose that $EMP(t)$, measured weekly, is the time-dependent variable being considered. An extended Cox model that does not allow for a lag-time assumes that the effect of EMP on the probability of survival at week t depends on the observed value of this variable at the same week t, and not, for example, at an earlier week. However, to allow for, say, a time-lag of one week, the EMP variable can be modified so that the hazard model at time t is predicted by EMP at week $t - 1$. Thus, the variable $EMP(t)$ is replaced in the model by the variable $EMP(t-1)$. We picture the difference in models as follows:

$$EMP(t) = \text{employment status at week } t.$$

Model without lag-time:

$$h(t|\underline{x}(t)) = h_0(t)\exp(\gamma EMP(t)).$$

$$\text{same week}$$

Model with 1-week lag-time:

$$h(t|\underline{x}(t)) = h_0(t)\exp(\gamma^* EMP(t-1)).$$

$$\text{one-week earlier}$$

Let L_j denote the lag-time specified for the time-dependent variable $X^{(j)}$. Then we can write the **general lag-time extended model** as:

$$h(t|\underline{x}(t)) = h_0(t)\exp\left(\sum_{i=1}^{p_1}\beta_i x^{(i)} + \sum_{j=1}^{p_2}\gamma_j x^{(j)}(t-L_j)\right).$$

$$x^{(j)}(t-L_j) \text{ replaces } x^{(j)}(t)$$

The estimated general hazard ratio (HR) formula for the extended Cox model is shown below. The most important feature of this formula is that the PH assumption is no longer satisfied. This formula describes the ratio of hazards at a particular time t, and requires the specification of two sets of predictors at time t. Denote these two sets by $\underline{x}^*$ and $\underline{x}$.

$$\widehat{\text{HR}}(t) = \frac{\hat{h}(t|\underline{x}^*(t))}{\hat{h}(t|\underline{x}(t))}$$

$$= \exp\left(\sum_{i=1}^{p_1}\hat{\beta}_i\left(x^{*(i)} - x^{(i)}\right) + \sum_{j=1}^{p_2}\hat{\gamma}_j\left(x^{*(j)}(t) - x^{(j)}(t)\right)\right).$$

Examples:

Example 1:

Consider the model with one time-independent exposure status variable E and one time-dependent predictor $x(t) = E \times t$. The extended Cox model is

$$h(t|\underline{x}(t)) = h_0(t)\exp\left(\beta E + \gamma(E \times t)\right).$$

The reduced model under $H_0: \gamma = 0$ contains only the E variable and hence PH is satisfied. However, when $\gamma \neq 0$, it is not. To compare exposed persons ($E = 1$) with unexposed persons ($E = 0$), we have

$$\underline{x}^*(t) = (E=1, E\times t=t),$$
$$\underline{x}(t) = (E=0, E\times t=0),$$

and $\quad\widehat{\text{HR}}(t) = \dfrac{\hat{h}(t|E=1, E\times t=t)}{\hat{h}(t|E=0, E\times t=0)}$

$$= \exp\left(\hat{\beta}(1-0) + \hat{\gamma}(t-0)\right)$$
$$= \exp\left(\hat{\beta} + \hat{\gamma}t\right).$$

If $\hat{\gamma} > 0$, the $\widehat{\text{HR}}(t)$ increases exponentially with time. The PH assumption is certainly not satisfied.

Note that the $\hat{\gamma}_j$ in the general formula is itself not time-dependent. Thus, this coefficient represents the **"overall"** **effect** of the corresponding time-dependent variable $X^{(j)}(t)$, considering all times at which this variable has been measured in the study.

Example 2:

Again we consider a one variable model. Let $E(t)$ denote a weekly measure of chemical exposure status at time t. That is, $E(t) = 1$ if exposed, and $= 0$ if unexposed, at a weekly measurement. As defined, the variable $E(t)$ can take on different patterns of values for different subjects. For example, for a five-week period, subject A's values may be 01011, whereas subject B's values may be 11010. In this model, the values of $E(t)$ may change over time for different subjects, but there is only one coefficient γ corresponding to the one variable in the model. The γ coefficient represents the overall effect on survival of the time-dependent variable $E(t)$. We picture this as:

$$
\begin{aligned}
E(t) \quad &= \quad \text{chemical exposure status at time } t \text{ (weekly)} \\
&= \quad \begin{cases} 0 & \text{if unexposed at time } t \\ 1 & \text{if exposed at time } t. \end{cases}
\end{aligned}
$$

A :
$$
\begin{array}{c|ccccc} E(t) & 0&1&0&1&1 \\ \hline t & 1&2&3&4&5 \end{array}
$$

B :
$$
\begin{array}{c|ccccc} E(t) & 1&1&0&1&0 \\ \hline t & 1&2&3&4&5 \end{array}
$$

$$h(t|\underline{x}(t)) \quad = \quad h_0(t)\exp(\gamma E(t)).$$

one coefficient γ
which represents the overall effect of $E(t)$.

The hazard ratio formula, which first compares an exposed person to an unexposed person at time t_1 and then compares the same two people at a later time t_2 where they are now both exposed, yields estimated HR's

$$
\begin{aligned}
\widehat{\text{HR}}(t_1) \quad &= \quad \frac{\hat{h}(t_1|E(t_1) = 1)}{\hat{h}(t_1|E(t_1) = 0)} \\
&= \quad \exp\left(\hat{\gamma}(1-0)\right) \\
&= \quad e^{\hat{\gamma}}, \text{ a constant,} \quad \text{and}
\end{aligned}
$$

$$\widehat{\text{HR}}(t_2) = \frac{\hat{h}(t_2|E(t_2)=1)}{\hat{h}(t_2|E(t_2)=1)}$$
$$= \exp(\hat{\gamma}(1-1))$$
$$= e^0 = 1, \text{ a different constant.}$$

This example illustrates that the hazard ratio $\widehat{\text{HR}}(t)$ is time-dependent. Thus, the PH assumption is not satisfied.

Example 3:

Let us again consider a model with the exposure variable E in Example 1 along with the variable $x(t) = E \times$ the "heavyside" function $g(t)$ where

$$g(t) = \begin{cases} 1 & \text{if } t \geq t_0 \\ 0 & \text{if } t < t_0. \end{cases} \tag{7.1}$$

The extended Cox model is

$$h(t|\underline{x}(t)) = h_0(t)\exp\left(\beta E + \gamma\left(E \times g(t)\right)\right)$$

and the hazard ratio is

$$\text{HR} = \frac{h(t|E=1)}{h(t|E=0)} = \exp\left(\beta + \gamma \times g(t)\right) = \begin{cases} \exp(\beta+\gamma) & \text{if } t \geq t_0 \\ \exp(\beta) & \text{if } t < t_0. \end{cases}$$

This hazard ratio has two distinct values. It differs over the two fixed intervals, but is constant over each individual interval.

An alternate form of this model is:

$$h(t|\underline{x}(t)) = h_0(t)\exp\left(\gamma_1\left(E \times g_1(t)\right) + \gamma_2\left(E \times g_2(t)\right)\right),$$

where

$$g_1(t) = \begin{cases} 1 & \text{if } t < t_0 \\ 0 & \text{if } t \geq t_0 \end{cases} \quad \text{and} \quad g_2(t) = \begin{cases} 1 & \text{if } t \geq t_0 \\ 0 & \text{if } t < t_0. \end{cases} \tag{7.2}$$

The hazard ratio is now expressed as

$$\text{HR} = \frac{h(t|E=1)}{h(t|E=0)} = \exp\left(\gamma_1 \times g_1(t) + \gamma_2 \times g_2(t)\right) = \begin{cases} \exp(\gamma_2) & \text{if } t \geq t_0 \\ \exp(\gamma_1) & \text{if } t < t_0. \end{cases} \tag{7.3}$$

For $\gamma < 0$, the graph of this HR is displayed in Figure 7.1.

Treatment of heroin addicts example:

The case study here explores an epidemiologic study on the treatment of heroin addicts. The data comes from a 1991 Australian study conducted by J. Caplehorn, *et al.* (1991). Kleinbaum (1996, pages 229–235) discusses this data. We present a shortened version of his analysis. The primary purpose of the Australian study was to compare the retention time in two methadone

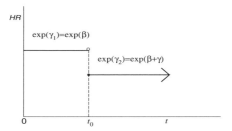

Figure 7.1 *The hazard ratio when the heavyside function is used.*

treatment clinics for heroin addicts. A patient's survival time T was determined as the time in days until s/he dropped out of the clinic or was censored at the end of the study. The two clinics differed according to their overall treatment policies. Did these two overall treatment policies significantly affect survival and was there a significant difference between the two effects? The primary exposure variable of interest is the clinic variable, which is coded as 1 or 2. The data frame is called ADDICTS. A summary of this data set follows.

Australian study of heroin addicts, Caplehorn, *et al.* (1991)

- two methadone treatment clinics
- T = days remaining in treatment
  (= days until drop out of clinic)
- clinics differ in overall treatment policies
- 238 patients in the study

Table 7.1: *Description of ADDICTS data set*

Data set: ADDICTS	
Column 1:	Subject ID
Column 2:	Clinic (1 or 2) ← **exposure variable**
Column 3:	Survival status
	0 = censored
	1 = departed clinic
Column 4:	Survival time in days
Column 5:	Prison record ← **covariate**
	0 = none, 1 = any
Column 6:	Maximum methadone dose (mg/day)← **covariate**

Part I:

The following is S code, along with modified output, that fits two K-M curves

Table 7.2:   *A compressed ADDICTS data set*

ID	Clinic	Status	Days.survival	Prison	Dose
1	1	1	428	0	50
2	1	1	275	1	55
3	1	1	262	0	55
⋮	⋮	⋮	⋮	⋮	⋮
263	2	0	551	0	65
264	1	1	90	0	40
266	1	1	47	0	45

not adjusted for any covariates to the survival data, conducts a log-rank test for significant differences between the two clinics' survival curves, and then plots the two curves.

```
> addict.fit <- survfit(Surv(Days.survival,Status)~Clinic,
                         data = ADDICTS)
> addict.fit
            n  events mean se(mean) median 0.95LCL 0.95UCL
Clinic=1  163     122  432     22.4    428     348     514
Clinic=2   75      28  732     50.5     NA     661      NA
> survdiff(Surv(Days.survival,Status)~Clinic,data = ADDICTS)
            N      Observed Expected (O-E)^2/E (O-E)^2/V
Clinic=1  163           122     90.9      10.6      27.9
Clinic=2   75            28     59.1      16.4      27.9
Chisq= 27.9  on 1 degrees of freedom, p= 1.28e-007
> plot(addict.fit, lwd = 3,col = 1,type = "l",lty=c(1, 3),
    cex=2,lab=c(10,10,7),xlab ="Retention time (days) in
    methadone treatment",ylab="Percent Retained",yscale=100)
```

Results:

- The log-rank test is highly significant with p-value= 1.28×10^{-7}.
- The graph in Figure 7.2 glaringly confirms this difference.
- This graph shows curve for clinic 2 is always above curve for clinic 1.
- Curves diverge, with clinic 2 being dramatically better after about one year in retention of patients in its treatment program.
- Lastly, this suggests the PH assumption is not satisfied.

Part II:

We now fit a Cox PH model which adjusts for the three predictor variables. This hazard model is

$$h(t|\underline{x}) = h_0 \exp(\beta_1 \text{Clinic} + \beta_2 \text{Prison} + \beta_3 \text{Dose}).$$

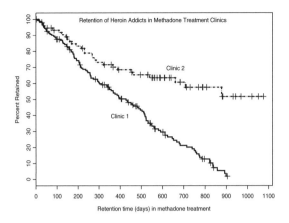

Figure 7.2 *K-M curves for ADDICTS not adjusted for covariates.*

The Grambsch-Therneau test of the PH assumption based on the scaled Schoenfeld residuals discussed in Chapter 6.3.5 is conducted. Recall the S function coxph.zph provides this test along with a plot of each component of $\widehat{\underline{\beta}}(t)$ against ordered time. A summary of the S output is:

```
> fit1 <- coxph(Surv(Days.survival,Status) ~ Clinic+Prison+
        Dose,data = ADDICTS,x = T) # Fits a  Cox PH model
> fit1
          coef   exp(coef)   se(coef)      z       p
Clinic  -1.0098    0.364     0.21488    -4.70   2.6e-006
Prison   0.3265    1.386     0.16722     1.95   5.1e-002
Dose    -0.0354    0.965     0.00638    -5.54   2.9e-008

Likelihood ratio test=64.6  on 3 df, p=6.23e-014  n= 238
> testph <- cox.zph(fit1)   # Tests the proportional
                            # hazards assumption
> print(testph) # Prints the results
           rho    chisq       p
Clinic  -0.2578   11.19    0.000824
Prison  -0.0382    0.22    0.639324
Dose     0.0724    0.70    0.402755
GLOBAL      NA    12.62    0.005546
> par(mfrow = c(2, 2))
> plot(testph)  # Plots the scaled Schoenfeld residuals.
```

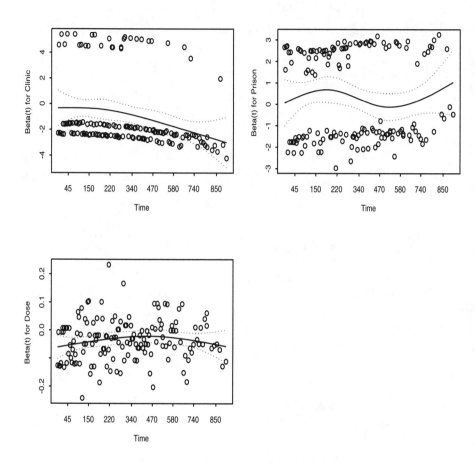

Figure 7.3 *Diagnostic plots of the constancy of the coefficients in the fit1 model. Each plot is of a component of $\widehat{\beta}(t)$ against ordered time. A spline smoother is shown, together with ± 2 standard deviation bands.*

Results:

- The GLOBAL test (a LRT) for non-PH is highly statistically significant with p-value = 0.005546.

- The p-values for Prison and Dose are very large, supporting that these variables are time-independent.

- The Grambsch-Therneau test has a p-value = 0.000824 for Clinic. This provides strong evidence that the variable Clinic violates the PH assumption and confirms what the graph in Figure 7.2 suggests.

- The plot of $\hat{\beta}_1(t)$, the coefficient for Clinic, against ordered time in Figure 7.3 provides further supporting evidence of this violation.

- We recommend finding a function $g(t)$ to multiply Clinic by; that is, create a defined time-dependent variable, and then fit an extended Cox model.

Part III: Stratified Cox model

We now stratify on the exposure variable Clinic and fit a Cox PH model to adjust for the two time-independent covariates Prison and Dose. Modified S output and a plot of the two adjusted K-M survival curves follow.

```
> fit2 <- coxph(Surv(Days.survival,Status) ~ strata(Clinic)+
                Prison+Dose,data=ADDICTS)
> fit2
            coef   exp(coef)  se(coef)      z       p
Prison    0.3896      1.476    0.16893    2.31   2.1e-002
Dose     -0.0351      0.965    0.00646   -5.43   5.6e-008

Likelihood ratio test=33.9  on 2 df, p=4.32e-008  n= 238
> survfit(fit2)
              n    events   mean  se(mean)  median  .95LCL  .95UCL
Clinic=1    162     122     434     22.0      434     358     517
Clinic=2     74      28     624     38.1      878     661      NA
# Note that each stratum has one less observation.
# This tells us that the shortest observed retention
# time in each clinic is censored.
> plot(survfit(fit2),lwd=3,col=1,type="l",lty=c(1,3),
    cex=2,lab=c(10,10,7),xlab="Retention time (days) in
    methadone treatment",ylab="Percent Retained",yscale=100)
> abline(v = 366,lty=3,lwd=2)
```

Results:

- Figure 7.4 provides same pictorial evidence as Figure 7.2; that is, curve for clinic 2 is always above clinic 1's curve, with clinic 2 being dramatically better in retention of patients in its treatment program after about one year.

- The estimated coefficients for Prison and Dose do not change much. This gives good evidence that the stratified model does satisfy the PH assumtion; hence, this analysis is valid.

- Figure 7.4 provides a picture of the effect of Clinic on retention of patients. But by stratifying on Clinic, we get no estimate of its effect; i.e., no estimated β_1 coefficient. Hence, we cannot obtain a hazard ratio for Clinic.

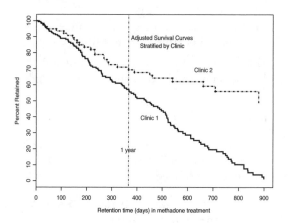

Figure 7.4 *K-M curves adjusted for covariates Prison and Dose, stratified by Clinic.*

- The exposure variable Clinic must be in the model in order to obtain a hazard for it. For this reason, we look now to the extended Cox model.

Part IV: An extended Cox model analysis

Here we use a model that contains two heavyside functions (7.2), $g_1(t)$ and $g_2(t)$, where $t_0 = 365$ days. Here the hazard model is

$$h(t|\underline{x}(t)) = h_0(t) \exp\left(\beta_1 \text{Prison} + \beta_2 \text{Dose} + \gamma_1 \left(\text{Clinic} \times g_1(t)\right)\right.$$
$$\left. + \gamma_2 \left(\text{Clinic} \times g_2(t)\right)\right)$$

where

$$g_1(t) = \left\{ \begin{array}{ll} 1 & \text{if } t < 365 \\ 0 & \text{if } t \geq 365 \end{array} \right. \qquad g_2(t) = \left\{ \begin{array}{ll} 1 & \text{if } t \geq 365 \\ 0 & \text{if } t < 365 \end{array} \right.$$

and

$$\text{Clinic} = \left\{ \begin{array}{ll} 1 & \text{if Clinic=1} \\ 0 & \text{if Clinic=2.} \end{array} \right. \qquad (7.4)$$

The hazard ratio for the exposure variable Clinic now varies with time. It assumes two distinct values depending whether time < 365 days or time ≥ 365 days. The form of the HR (7.3) is

$$t < 365 : \quad \text{HR} = \exp(\gamma_1)$$
$$t \geq 365 : \quad \text{HR} = \exp(\gamma_2).$$

Time-dependent covariates effect the rate for upcoming events. In order to implement an extended Cox model properly in S using the coxph function, one must use the Anderson-Gill (1982) formulation of the proportional hazards model as a counting process. They treat each subject as a very slow Poisson process. A censored subject is not viewed as incomplete, but as one whose event count is still zero. For a brief introduction to the counting process approach, see Appendix 2 of Hosmer & Lemeshow (1999) and the on-line manual S-PLUS 2000, Guide to Statistics, Vol 2, Chapter 10. Klein & Moeschberger (1997, pages 70−79) discuss this counting process formulation. They devote their Chapter 9 to the topic of modelling time-dependent covariates. For a more advanced and thorough treatment of counting processes in survival analysis, see Fleming and Harrington (1991).

The ADDICTS data set must be reformulated to match the Anderson-Gill notation. To illustrate this, consider the following cases: In both cases the t denotes the patient's recorded survival time, whether censored or not.

Case 1: For $t < 365$, $g_1(t) = 1$ and $g_2(t) = 0$. Here a patient's data record is just one row and looks like this:

Start	Stop	Status	Dose	Prison	Clinicg1t	Clinicg2t
0	t	same	same	same	Clinic	0

Case 2: For $t \geq 365$, $g_1(t) = 0$ and $g_2(t) = 1$. Here a patient's data record is formulated into two rows and looks like this:

Start	Stop	Status	Dose	Prison	Clinicg1t	Clinicg2t
0	365	0	same	same	Clinic	0
365	t	same	same	same	0	Clinic

The following S program puts the ADDICTS data set into the counting process form and then fits the extended Cox model stated above. We employ the function extcox.1Et(DATA,t_0). The arguments are: DATA: a data frame; t_0: specified time for the heavyside functions $g_1(t)$ and $g_2(t)$. One must put the exposure variable and the time variable into columns 2 and 4, respectively. Although this is designed for one exposure variable, it can be easily extended to more than one. The output from coxph has been modified. From expression (7.4), t_0 is set to 365.

```
> out <- extcox.1Et(ADDICTS,365)
  # The fitted extended Cox model is:
> Clinicg1t <- out$ET1 # Exposure x g1(t)
> Clinicg2t <- out$ET2 # Exposure x g2(t)
> fit3 <- coxph(Surv(Start,Stop,Status) ~ Prison+Dose+
              Clinicg1t+Clinicg2t,data=out)
```

Table 7.3: *Summary of the fitted extended Cox model*

	coef	HR exp(coef)	se(coef)	z	p
Prison	0.3779	1.459	0.16841	2.24	2.5e-002
Dose	-0.0355	0.965	0.00643	-5.51	3.5e-008
Clinicg1t	0.4594	1.583	0.25529	1.80	7.2e-002
Clinicg2t	1.8284	6.224	0.38567	4.74	2.1e-006

Likelihood ratio test=74.2 on 4 df, p=2.89e-015 n=360

Table 7.4: *95% Confidence intervals for Clinic's HR's*

95% C.I.'s for the Clinic's HR's	
$t < 365$:	[0.960, 2.611]
$t \geq 365$:	[2.922, 13.254]

Results:

- Table 7.3 shows a borderline nonsignificant $\widehat{HR} = 1.583$ with p-value = 0.072 for the effect of Clinic when time < 365 days. But for $t \geq 365$, the $\widehat{HR} = 6.224$ is highly significant with p-value = 2.1×10^{-6}.

- Table 7.4 reports confidence intervals for the two HR's. The general form of these 95% C.I.'s is $\exp(\text{coef} \pm 1.96 \times \text{se(coef)})$. The 95% C.I. for the HR when t precedes 365 covers 1 and is narrow. This supports a chance effect due to clinic during the first year. The 95% C.I. for the HR when $t \geq 365$ lies above 1 and is very wide showing a lack of precision.

- These findings support what was displayed in Figure 7.4. There is strong statistical evidence of a large difference in clinic survival times after one year in contrast to a small and probably insignificant difference in clinic survival times prior to one year, with clinic 2 always doing better in retention of patients than clinic 1. After one year, clinic 2 is 6.224 times more likely to retain a patient longer than clinic 1. Also, clinic 2 has $\frac{1}{6.224} \approx 16\%$ the risk of clinic 1 of a patient leaving its methadone treatment program.

- See Kleinbaum (1996, Chapter 6) for further analysis of this data.

- An alternative regression quantile analysis as presented in Chapter 8 may be appropriate when the PH assumption seems to be violated.

7.2 Competing risks: cumulative incidence estimator

Objectives of this section:

After studying Section 7.2, the student should:

1. Know the definition of a **competing risk** and give examples.
2. Understand why $1 - KM$ fails as an estimate of failure probabilities when a competing risk is present.
3. Understand what the **observable quantities** are.
4. Know the quantity referred to as the **cumulative incidence function**.
5. Know the definition of the **cause-specific hazard function**.
6. Know the definition of the **cumulative incidence (CI) estimator** and how to compute it in S.
7. Understand why the CI estimator is the appropriate quantity to use in the presence of a competing risk when estimation of failure probabilities from the event of interest by time t is the goal.

The material presented here is an adaptation of Gooley *et al.* (1999, 2000) and of material found in Kalbfleisch & Prentice (1980), Chapter 7, pages 163–169. Here we use KM to denote the K-M estimator of survival.

A **competing risk** is defined as an event whose occurrence either precludes the occurrence of the event under examination or fundamentally alters the probability of occurrence of this event of interest.

Examples:

1. The cohort under study is a group of patients all diagnosed with insulin dependent diabetes mellitus (IDDM). The outcome of interest is the occurrence of end-stage renal failure. Let T = time from diagnosis of IDDM to end-stage renal failure. We wish to estimate the probability of developing end-stage renal failure by time t; i.e., $P(T \le t)$. A competing risk could be death without end-stage renal failure. If a patient with IDDM dies without renal failure, it is impossible for the patient to experience the outcome of interest – time to end-stage renal failure.

2. A potential complication among patients who receive a bone marrow transplant is known as chronic graft-versus-host disease (CGVHD), and the probability of this complication is often an outcome of interest in clinical trials. For this example we take the occurrence of CGVHD to be the failure of interest. Competing risks for the occurrence of CGVHD are death without CGVHD and relapse without CGVHD. Relapse is considered to be a competing risk because patients who experience a relapse of their disease are often withdrawn from the immunosuppressive therapy, where this therapy is given primarily as prophylaxis for development of CGVHD. Relapse

therefore fundamentally alters the probability of occurrence of CGVHD and for this reason is regarded as a competing risk.

When all deaths can be attributed to a single "cause of interest," the KM method gives a consistent estimate of the survival function in the presence of right censoring. However, when patients leave the study (die or are withdrawn) for other causes, the KM estimate becomes biased. That is, it fails to handle competing-risk deaths appropriately. In this section we introduce the cumulative incidence (CI) estimator, which continues to be consistent when competing risks as well as random censoring occurs.

In the presence of competing risks three mutually exclusive outcomes are possible for each patient under study:

1. fail from event of interest,

2. fail from competing risk, or

3. survive without failure to last contact (censored).

$1-$KM does not handle failures from a competing risk in its calculation as it treats competing-risk failures as censored. But this is wrong. What's the difference?

- Censored patients (not failed by last contact) still have the potential to fail.

- Patients who fail from a competing risk no longer have the potential to fail from the event of interest.

As $1-$KM treats competing-risk failures as censored, we learn from the redistribute-to-the-right algorithm detailed in Table 2.2 of Chapter 2.1 that failures occurring after a competing-risk failure contribute more than is appropriate. This overestimates the probability of failure from the event of interest at all times after the competing-risk time. This is a result of redistributing to the right a potential amount that does not exist.

Let's consider the following hypothetical data where z_i, $i = 1, \ldots, 5$ denotes the distinct times, "$+$" denotes a censored time, and "$\cdot$" denotes a death time from cause of interest. There are a total of 7 observed values. On a time line we have

Suppose the two censored times at z_3 are now competing-risk failures. Following the redistribute-to-the-right algorithm, the $1-$KM estimate, for example at z_4, is $\frac{5}{7}$. But this is inappropriate. All 7 patients have complete follow-up to time z_4. This means each has either

1. experienced the event of interest— there are 4,

2. failed from a competing risk— there are 2, or

3. is still alive (survived beyond z_4)— there is 1.

Therefore, the estimate of probability of failure from the event of interest by time z_4 is $\frac{4}{7}$. The CI estimate does give this correct value $\frac{4}{7}$.

More formally, let T_1 denote the time for the event of interest (type 1 failure) to occur. Let T_2, which is assumed to be independent of T_1, denote the time for a competing risk (type 2 failure) to occur. Although we are interested in estimating the probability of type 1 failure by time t, we can only observe T_{i1}, a type 1 failure time for the ith patient, when $T_{i1} < T_{i2}, i = 1, \ldots, n$. Hence, the **estimable quantity** of interest is

$$P(T_1 \leq t \text{ and } T_1 < T_2). \tag{7.5}$$

This quantity increases to $P(T_1 < T_2)$ as t increases to ∞. When there are no censored values present, this quantity is directly estimated by the observed proportion

$$\frac{\# \text{ of observed type 1 failure times} \leq t}{n}.$$

Let $T_M = \min\{T_1, T_2\}$ and let ξ denote the cause-specific failure of either type 1 or type 2. Then $\xi = 1$ or 2. The **observable event** $\{T_1 \leq t \text{ and } T_1 < T_2\}$ is equivalently expressed by the event $\{T_M \leq t \text{ and } \xi = 1\}$. Let $F_1(t)$ denote the probability of this event. Then

$$F_1(t) = P(T_1 \leq t \text{ and } T_1 < T_2) = P(T_M \leq t \text{ and } \xi = 1). \tag{7.6}$$

Kalbfleisch and Prentice (1980, page 168) name $F_1(t)$, a subdistribution function, the **cumulative incidence function**.

We now derive an expression for $F_1(t)$ which lends itself to be easily estimated by a function of known estimators that accommodate right-censored observations. Following the definition of hazard function on page 6, the **cause-specific hazard function** for a type 1 failure by time t is defined to be

$$
\begin{aligned}
h_1(t) &= \lim_{\Delta t \to 0} \frac{P(t \leq T_M < t + \Delta t, \xi = 1 | T_M > t)}{\Delta t} \\
&= \lim_{\Delta t \to 0} \frac{P(t < T_M < t + \Delta t, \xi = 1)}{\Delta t} \times \frac{1}{S_M(t)} \\
&= \frac{f_1(t)}{S_M(t)},
\end{aligned}
$$

where $S_M(t)$ denotes the survival function of T_M and $f_1(t)$ is the subdensity corresponding to $F_1(t)$. Let $H_1(t)$ denote the type 1 cause-specific cumulative hazard function. Then

$$
\begin{aligned}
F_1(t) &= P(T_M \leq t, \xi = 1) \\
&= \int_0^t f_1(u)du \\
&= \int_0^t S_M(u)h_1(u)du \\
&= \int_0^t S_M(u)dH_1(u).
\end{aligned}
$$

Thus, when censored values are present, we use the Nelson-Aalen estimator (2.9) for $H_1(t)$ along with an appropriate empirical estimate of $S_M(t)$.

To construct this estimate we assume the n patients under study will experience one and only one of three distinct outcomes and the times at which these outcomes occur are ordered $t_1 \leq \ldots \leq t_n$. Let z_j denote the distinct ordered observed times, $j = 1, \ldots, m \leq n$. The three potential outcomes for a patient at time z_j are

1. type 1: patient fails from event of interest at z_j,

2. type 2: patient fails from a competing risk at z_j, or

3. censored: patient has not failed from either cause, but has follow-up only to z_j.

Let
$n =$ total # of patients under study; # initially at risk,
$e_j =$ # of patients who have type 1 failure at z_j,
$r_j =$ # of patients who have type 2 failure at z_j,
$c_j =$ # of patients who are censored at z_j,
$n_j^* =$ # of patients who are at risk of failure beyond (to the right of) z_j, and
$n_{j-1}^* =$ # of patients who are at risk just before z_j.

Then $n_j^* = n - \sum_{k=1}^{j}(e_k + r_k + c_k)$.

The KM estimate of survival without failure of type 1 is

$$
\mathrm{KM}_1(t) = \prod_{z_j \leq t} \left(\frac{n_{j-1}^* - e_j}{n_{j-1}^*} \right).
$$

Note that:

1. $\mathrm{KM}_1(t)$ is calculated by censoring the patients who fail from a competing risk.

2. $\frac{e_j}{n_{j-1}^*}$ estimates the hazard of type 1 failure at time z_j.

3. $\mathrm{KM}_1(t)$ does not depend on hazard of type 2 failure.

The KM estimate of survival without failure of type 2 is

$$\mathrm{KM}_2(t) = \prod_{z_j \leq t} \left(\frac{n_{j-1}^* - r_j}{n_{j-1}^*} \right).$$

An overall KM survivor function, $\mathrm{KM}_{12}(t)$, is obtained by considering failures of any type as events and represents the probability of surviving all causes of failure beyond time t. Define $\mathrm{KM}(t)_{12}$ as

$$\mathrm{KM}_{12}(t) = \mathrm{KM}_1(t) \times \mathrm{KM}_2(t).$$

This estimates the quantity $S_M(t) = P(\min\{T_1, T_2\} > t)$. The estimate $\mathrm{KM}_{12}(t)$ does depend on the hazard of each failure type.

Replacing $S_M(u)$ with the $\mathrm{KM}_{12}(u)$ estimate along with the Nelson–Aalen estimate $\widetilde{H}_1(u) = \sum_{z_j \leq u} \frac{e_j}{n_{j-1}^*}$ yields the **CI estimator** defined by Kalbfleisch and Prentice (1980, page 169). More precisely, let $\mathrm{CI}(t)$ denote the CI estimator. Then

$$\mathrm{CI}(t) = \sum_{z_j \leq t} \frac{e_j}{n_{j-1}^*} \times \mathrm{KM}_{12}(z_j). \tag{7.7}$$

Note that the CI estimator depends on estimates of the hazard of each failure type.

Now, any appropriate estimator for the CI function should change when and only when a patient fails from the event of interest (type 1). Each time a failure occurs the estimate should increase (jump) by a specified amount. Otherwise, there should be no jump. Hence, an appropriate estimate of the probability of type 1 failure by time t can be alternatively represented as

$$\mathrm{I}(t) = \sum_{z_j \leq t} J(z_j) e_j,$$

where $J(z_j)$ represents a patient's contribution to the "jump" in the estimate at time z_j. As will be seen, the "jump" sizes differ for 1–KM and CI because of the way each is calculated. We exploit the redistribute-to-the-right algorithm to give an equivalent representation for CI and KM which makes them easy to compute and elucidates why the CI estimator provides an appropriate estimate of the probability of failure from the event of interest by time t.

Let $J_{\mathrm{CI}}(t)$ and $J_{\mathrm{M}}(t)$ denote the $J(t)$ for the CI and KM estimates, respectively. Recall that the potential contribution to the estimate of probability of failure by time t from the event of interest for censored patients is equally redistributed among all patients at risk of failure beyond time t. Initially, all patients are equally likely; and hence, each has potential contribution $1/n$ to the estimate. Therefore

$$J_{\mathrm{CI}}(z_1) = \frac{1}{n}.$$

If censored, redistribute to the right equally the amount $(1/n) \times c_1$. Now, let

$J_{CI}(z_j)$ equal an individual's potential contribution to the estimate of probability of failure from the event of interest at time z_j. If censored, this amount is equally redistributed to all patients to the right. There are n_j^* of them. Hence,

$$\frac{J_{CI}(z_j)}{n_j^*} \times c_j = \text{additional contribution given to each to the right.}$$

Therefore, each of the n_j^* patients' potential contribution at time point z_{j+1} is

$$J_{CI}(z_{j+1}) = J_{CI}(z_j) + \frac{J_{CI}(z_j)}{n_j^*} \times c_j$$
$$= J_{CI}(z_j)\left(1 + \frac{c_j}{n_j^*}\right),$$

where $j = 0, \ldots, m \leq n - 1$, and $J_{CI}(z_0) = J_{CI}(z_1) = 1/n$ with $z_0 = 0$ and $c_0 = 0$. It can be shown the **CI estimator** defined by Kalbfleisch and Prentice is precisely expressed as

$$CI(t) = \sum_{z_j \leq t} J_{CI}(z_j)\left(1 + \frac{c_j}{n_j^*}\right) e_{j+1},$$

where $j = 0 \ldots, m \leq n - 1$. Note that if $e_j = 0$, the patients at z_j are either censored or have failure type 2 and $CI(t)$ will not increase. If a patient dies from a competing risk at z_j, his contribution is not redistributed and his contribution becomes zero beyond time z_j.

Following from Table 2.2 of Chapter 2.1, the $1-$KM estimate at time point z_{j+1} calculates each of the n_j^* individuals' potential contribution to the jump by

$$J_{KM}(z_{j+1}) = J_{KM}(z_j)\left(1 + \frac{(c_j + rj)}{n_j^*}\right)$$

and thus the $1-$KM estimate is reexpressed as

$$1 - KM(t) = \sum_{z_j \leq t} J_{KM}(z_j)\left(1 + \frac{(c_j + r_j)}{n_j^*}\right) e_{j+1}.$$

Here we see precisely why $1-$KM fails to be an appropriate estimate of the probability of failure from the event of interest by time t. It treats a competing-risk failure as censored and, hence, redistributes its potential contribution to the right. In the calculation of the CI, patients who die from a competing risk are correctly assumed to be unable to fail from the event of interest beyond the time of the competing-risk failure. They have no potential to redistribute-to-the-right. If there are NO failures from a competing risk,

$$1 - KM = CI.$$

But if competing-risk failures in the observed times exist, $1-$KM $>$ CI at and beyond the time of the first failure from event of interest that follows a competing-risk failure.

We illustrate the calculation of both estimators in Table 7.5. We consider the following hypothetical data of eight observations where an "r" denotes a competing-risk failure and "$+$" denotes a censored observation. There are $m = 6$ distinct times, two censored times, and one competing-risk time.

On a time line we have

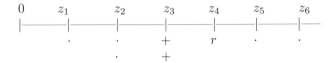

Table 7.5: *Comparison of* CI(t) *and* $1-$KM(t) *in the presence of a competing risk*

z_j	n_j^*	e_j	c_j	r_j	$J_{\mathrm{CI}}(z_{j+1})$	CI(t)	$J_{\mathrm{KM}}(z_{j+1})$	$1-$KM(t)
0	8	0	0	0	$\frac{1}{8}$	0	$\frac{1}{8}$	0
z_1	7	1	0	0	$\frac{1}{8}\left(1+\frac{0}{8}\right)=\frac{1}{8}$	$\frac{1}{8}$	$\frac{1}{8}\left(1+\frac{0}{8}\right)=\frac{1}{8}$	$\frac{1}{8}$
z_2	5	2	0	0	$\frac{1}{8}\left(1+\frac{0}{7}\right)=\frac{1}{8}$	$\frac{3}{8}$	$\frac{1}{8}\left(1+\frac{0}{7}\right)=\frac{1}{8}$	$\frac{3}{8}$
z_3	3	0	2	0	$\frac{1}{8}\left(1+\frac{0}{5}\right)=\frac{1}{8}$	$\frac{3}{8}$	$\frac{1}{8}\left(1+\frac{0}{5}\right)=\frac{1}{8}$	$\frac{3}{8}$
z_4	2	0	0	1	$\frac{1}{8}\left(1+\frac{2}{3}\right)=\frac{5}{24}$	$\frac{3}{8}$	$\frac{1}{8}\left(1+\frac{2}{3}\right)=\frac{5}{24}$	$\frac{3}{8}$
z_5	1	1	0	0	$\frac{5}{24}\left(1+\frac{0}{2}\right)=\frac{5}{24}$	$\frac{7}{12}$	$\frac{5}{24}\left(1+\frac{1}{2}\right)=\frac{5}{16}$	$\frac{11}{16}$
z_6	0	1	0	0	$\frac{5}{24}\left(1+\frac{0}{1}\right)=\frac{5}{24}$	$\frac{19}{24}$	$\frac{5}{16}\left(1+\frac{0}{1}\right)=\frac{5}{16}$	1

Because of lack of book space we do not provide an example. However, Gray (2002) provides a competing risks library which provides functions to compute CI estimates along with estimated variances. The library is available on the web to download. It is called cmprsk and is designed for S-PLUS 3.4 unix platform, for S-PLUS 2000 with Windows 95 or later, and for R with Windows 95 or later. The library contains functions for analysis of subdistribution functions (7.5 or 7.6) in competing risks. The function cuminc() computes the CI(t) estimator (7.7), which estimates subdistributions and performs a nonparametric test for equality of subdistributions across groups. Ancillary functions print.cuminc() and plot.cuminc() are also included. For the S user, go to http://biowww.dfci.harvard.edu/~gray. Scroll downward. There you will find two versions of the cmprsk library. For the R user, go to http://www.r-project.org/. Then click on <u>CRAN</u> on the left. Click any site in the list; for example, ETH Zürich. Then $\rightarrow$ Windows (95 or later) $\rightarrow$ contrib $\rightarrow$ cmprsk.zip.

7.3 Analysis of left-truncated and right-censored data

According to Cox and Oakes (1984, page 2),

> The time origin need not always be at the point at which an individual enters the study, but if it is not, special methods are needed.

In this section we consider left-truncated and right-censored (LTRC) data. We describe how the K-M estimator (2.2) and its standard error (2.3) for right-censored data are extended to analyze LTRC data. We also describe how to extend the Cox PH model, discussed in Chapter 5, to analyze LTRC data.

Objectives of this section:

After studying Section 7.3, the student should:

1. Be able to identify LTRC data and know how to construct the likelihood function.
2. Know how to compute the **modified Kaplan-Meier (K-M) estimate** of survival and **modified Greenwood's estimate** of asymptotic variance of modified K-M at time t.
3. Know how to plot the modified K-M curve over time t in S.
4. Know how to implement the S function `survfit` to conduct nonparametric analysis.
5. Know how to plot two modified K-M curves to compare survival between two (treatment) groups.
6. Know how to implement the S function `coxph` to run Cox PH model for LTRC data.

Cox and Oakes (1984, pages 177−178) explain left truncation arises when individuals come under observation only some known time after the natural time origin of the phenomenon under study. That is, had the individual failed before the truncation time in question, that individual would not have been observed. Therefore, in particular, any contribution to the likelihood must be conditional on the truncation limit having been exceeded. If for the ith individual the left truncation time limit is x_i, possibly zero, and the individual either fails at time t_i or is right-censored at c_i, the contribution to the likelihood for a homogeneous sample of individuals is either

$$\frac{f(t_i)}{S_f(x_i)} \quad \text{or} \quad \frac{S_f(c_i)}{S_f(x_i)}.$$

Examples:

1. In epidemiological studies of the effects on mortality of occupational exposure to agents such as asbestos, the natural measure of time is age, since this is such a strong determinant of mortality. However, observation on each individual commences only when he/she starts work in a job which involves exposure to asbestos (Cox and Oakes, 1984).

2. In industrial reliability studies, some components may already have been in use for some period before observation begins (Cox and Oakes, 1984).

3. Left truncation occurs in astronomy often as one can only detect objects which are brighter than some limiting flux during an observing session with a given telescope and detector. The magnitude limit of the telescope and detector can be considered as a left truncation variable. There is a great need for statisticians to evaluate methods developed over decades by astronomers to treat truncated data (Babu and Feigelson, 1996).

4. Woolson (1981) reported survival data on 26 psychiatric inpatients admitted to the University of Iowa hospitals during the years 1935 – 1948. In this study, the time of interest is the death time of each psychiatric inpatient. To be included in the study, one must have experienced some sort of psychiatric symptom. One's age at admission to the hospital is one's left truncation time as anyone who has not experienced a psychiatric symptom will not be included in this study. This data set is analyzed throughout this section.

Each individual has a left truncation time (delayed entry time) X_i, a lifetime T_i, and a censoring time C_i. Denote $k(\cdot)$ a p.d.f. of X, $f(\cdot)$ and $S_f(\cdot)$ a p.d.f. and a survivor function of T, respectively, and $g(\cdot)$ and $S_g(\cdot)$ a p.d.f. and a survivor function of G, respectively. On each of n individuals we observe (X_i, Y_i, δ_i) where

$$Y_i = \min(T_i, C_i) \quad \text{and} \quad \delta_i = \begin{cases} 1 & \text{if} \quad X_i < T_i \le C_i \\ 0 & \text{if} \quad X_i < C_i < T_i \end{cases}.$$

Hence we observe n iid random triples (X_i, Y_i, δ_i). The times X_i, T_i, and C_i are assumed to be independent.

The conditional likelihood function of the n iid triples (X_i, Y_i, δ_i) given $T_i > X_i$ is given by

$$
\begin{aligned}
L &= \prod_{i=1}^{n} \left(\frac{k(x_i)f(y_i)S_g(y_i)}{S_f(x)} \right)^{\delta_i} \cdot \left(\frac{k(x_i)g(y_i)S_f(y_i)}{S_f(x)} \right)^{1-\delta_i} \\
&= \left(\prod_{i=1}^{n} k(x_i)S_g(y_i)^{\delta_i} g(y_i)^{1-\delta_i} \right) \cdot \left(\prod_{i=1}^{n} \frac{f(y_i)^{\delta_i} S_f(y_i)^{1-\delta_i}}{S_f(x)} \right).
\end{aligned}
$$

The derivation of the conditional likelihood is as follows:

$$
\begin{aligned}
P(X = x, Y = y, \delta = 0 | T > x) &= P(X = x, C = y, C < T | T > x) \\
&= \frac{P(X = x, C = y, y < T)}{S_f(x)} \\
\text{(by independence)} \quad &= \frac{k(x)P(C = y)P(y < T)}{S_f(x)} \\
&= \frac{k(x)g(y)S_f(y)}{S_f(x)}.
\end{aligned}
$$

$$P(X = x, Y = y, \delta = 1 | T > x) = \frac{P(X = x, T = y, T < C)}{S_f(x)}$$

$$= \frac{k(x)P(T = y, y < C)}{S_f(x)} = \frac{k(x)f(y)S_g(y)}{S_f(x)}.$$

Hence, the conditional p.d.f. of the triple (X, Y, δ) given $T > X$ (a mixed distribution as Y is continuous and δ is discrete) is given, up to a multiplicative constant, by the single expression

$$P(x, y, \delta) = \frac{f(y)^\delta S_f(y)^{1-\delta}}{S_f(x)} = \frac{h(y)^\delta S_f(y)}{S_f(x)}, \tag{7.8}$$

where $h(\cdot)$ is the hazard function of T.

Remarks:

1. When all the left truncation time limits $x_i's$ are zero, LTRC data become right-censored (RC) data.

2. Left truncation is substantially different from left censoring. In left censoring, we include in the study individuals who experience the event of interest prior to their entry times whereas in left truncation we do not.

3. Tsai (1990) considers a test for the independence of truncation time and failure time. He develops a statistic generalizing Kendall's tau and provides asymptotic properties of the test statistic under the independence assumption.

4. Efron and Petrosian (1992) also consider a test of independence for truncated data. They develop a truncated permutation test for independence and apply it to red shift surveys in astronomical studies.

5. Although we mainly focus on nonparametric methods of analysis, we briefly consider the exponential model and leave the details as an exercise. With $f(t; \lambda) = \lambda \exp(-\lambda t)$, each individual's contribution to the likelihood is either

$$\frac{f(t; \lambda)}{S_f(x; \lambda)} = \frac{\lambda \exp(-\lambda t)}{\exp(-\lambda x)} = \lambda \exp(-\lambda(t - x))$$

or

$$\frac{S_f(c; \lambda)}{S_f(x; \lambda)} = \frac{\exp(-\lambda c)}{\exp(-\lambda x)} = \exp(-\lambda(c - x)).$$

Then for a sample of n iid triples (X_i, Y_i, δ_i) from exponential distribution with the parameter λ, the joint likelihood function is given

$$L(\lambda) = \prod_{i=1}^{n} \Big(\lambda \exp(-\lambda(t_i - x_i)) \Big)^{\delta_i} \cdot \Big(\exp(-\lambda(c_i - x_i)) \Big)^{1-\delta_i}$$

$$= \lambda^{n_u} \exp\Big(-\lambda \sum_{i=1}^{n} (y_i - x_i) \Big), \tag{7.9}$$

where n_u is the number of uncensored observations. WHY! One can easily notice that this is almost identical to the joint likelihood for RC data from the exponential distribution with the parameter λ. The only difference is that we now have $y_i - x_i$ instead of y_i in the likelihood.

7.3.1 Modified Kaplan-Meier (K-M) estimator of the survivor function for LTRC data

Modified K-M estimator adjusts the K-M estimator to reflect the presence of left truncation.

This estimator along with a modified estimator of its variance was proposed by Tsai, Jewell, and Wang (1987). On each of n individuals we observe the triple (X_i, Y_i, δ_i) where

$$Y_i = \min(T_i, C_i) \text{ and } \delta_i = \begin{cases} 1 & \text{if } X_i < T_i \le C_i \\ 0 & \text{if } X_i < C_i < T_i. \end{cases}$$

Let $t_{(1)}, \ldots, t_{(r)}$ denote the $r \le n$ distinct ordered (uncensored) death times, so that $t_{(j)}$ is the jth ordered death time. We now define the **modified risk set** $\mathcal{R}(t_{(i)})$ at $t_{(i)}$ by

$$\mathcal{R}(t_{(i)}) = \{j | x_j \le t_{(i)} \le y_j\}, \quad j = 1, \ldots, n, \quad i = 1, \ldots, r, \quad (7.10)$$

where

$$
\begin{aligned}
n_i &= \ \# \text{ in } \mathcal{R}(t_{(i)}) \\
&= \ \# \text{ left-truncated before } t_{(i)} \text{ and alive (but not censored) just} \\
&\quad \text{before } t_{(i)} \\
d_i &= \ \# \text{ left-truncated before } t_{(i)} \text{ but died at time } t_{(i)} \\
p_i &= \ P(\text{surviving through } t_{(i)} \,|\, \text{left-truncated before } t_{(i)} \text{ and alive just} \\
&\quad \text{before } t_{(i)}) \\
q_i &= \ 1 - p_i.
\end{aligned}
$$

Notice the size of the risk set may not decrease strictly monotonically as the death time increases in contrast to the case of RC data. The estimates of p_i and q_i are

$$\widehat{q_i} = \frac{d_i}{n_i} \quad \text{and} \quad \widehat{p_i} = 1 - \widehat{q_i} = 1 - \frac{d_i}{n_i} = \left(\frac{n_i - d_i}{n_i} \right).$$

The **modified K-M estimator of the survivor function** is

$$\widehat{S}(t) = \prod_{t_{(i)} \le t} \widehat{p_i} = \prod_{t_{(i)} \le t} \left(\frac{n_i - d_i}{n_i} \right) = \prod_{i=1}^{k} \left(\frac{n_i - d_i}{n_i} \right), \quad (7.11)$$

where $t_{(k)} \le t < t_{(k+1)}$.

Psychiatric inpatients example:

Table 7.6 displays the psychiatric inpatients' data introduced on page 203. We use this data set to compute the modified K-M survival function estimate and the modified Greenwood's estimate of the variance of the survival function. We first modify Table 7.6 into the form of LTRC data, Table 7.7. Here Y equals Age at entry plus Yrs of follow-up. In the next section, we also use this data set to show how to run the Cox PH model for LTRC data. We now

Table 7.6: *Survival and demographic data for 26 psychiatric patients*

Sex*	Age at entry	Follow-up†	Sex*	Age at entry	Follow-up†
F	51	1	F	58	1
F	55	2	F	28	22
M	21	30+	M	19	28
F	25	32	F	48	11
F	47	14	F	25	36
F	31	31+	M	24	33+
M	25	33+	F	30	37+
F	33	35+	M	36	25
M	30	31+	M	41	22
F	43	26	F	45	24
F	35	35+	M	29	34+
M	35	30+	M	32	35
F	36	40	M	32	39+

*F = female, M = male; †Yrs from admission to death or censoring; + = censored (*Source*: Woolson (1981))

Table 7.7: *Survival data for psychiatric patients in LTRC format*

Sex	entry	futime	status	Sex	entry	futime	status
F	51	52	1	F	58	59	1
F	55	57	1	F	28	50	1
M	21	51	0	M	19	47	1
F	25	57	1	F	48	59	1
F	47	61	1	F	25	61	1
F	31	62	0	M	24	57	0
M	25	58	0	F	30	67	0
F	33	68	0	M	36	61	1
M	30	61	0	M	41	63	1
F	43	69	1	F	45	69	1
F	35	70	0	M	29	63	0
M	35	65	0	M	32	67	1
F	36	76	1	M	32	71	0

consider the psychiatric inpatients' data on a time line where an "x" denotes an entry time (age at entry), "•" a death time (age at entry plus years from admission to death), and "∘" a right-censored time (age at entry plus years from admission to censoring). Figure 7.5 graphically presents Table 7.7.

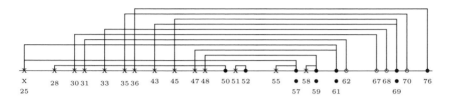

Figure 7.5 *Graphical presentation of LTRC data.*

We now compute the modified K-M estimate of the survivor function (7.11).

$$\widehat{S}(0) = 1$$
$$\widehat{S}(50) = \widehat{S}(0) \times \frac{12 - 1}{12} = .917$$
$$\widehat{S}(52) = \widehat{S}(50) \times \frac{12 - 1}{12} = .840. \qquad (7.12)$$

Note that the one who entered at the age of 28 but died at 50 is not included in the risk set $\mathcal{R}(52)$. The rest of the computation is left as Exercise 7.1.

Estimate of variance of $\widehat{S}(t)$: modified Greenwood's formula

$$\widehat{\mathrm{var}}\left(\widehat{S}(t)\right) = \widehat{S}^2(t) \sum_{t_{(i)} \leq t} \frac{d_i}{n_i(n_i - d_i)} = \widehat{S}^2(t) \sum_{i=1}^{k} \frac{d_i}{n_i(n_i - d_i)}, \qquad (7.13)$$

where $t_{(k)} \leq t < t_{(k+1)}$. Example with the female group in the psychi data:

$$\widehat{\mathrm{var}}\left(\widehat{S}(52)\right) = (.84)^2 \left(\frac{1}{12(12 - 1)} + \frac{1}{12(12 - 1)}\right) = .0107$$
$$\mathrm{s.e.}\left(\widehat{S}(52)\right) = .1034. \qquad (7.14)$$

The theory tells us that for each fixed value t

$$\widehat{S}(t) \overset{a}{\sim} \mathrm{normal}\left(S(t), \widehat{\mathrm{var}}\left(\widehat{S}(t)\right)\right).$$

Thus, at time t, an approximate $(1 - \alpha) \times 100\%$ confidence interval for the probability of survival, $S(t)=P(T > t)$, is given by

$$\widehat{S}(t) \pm \mathrm{z}_{\frac{\alpha}{2}} \times \mathrm{s.e.}\left(\widehat{S}(t)\right), \qquad (7.15)$$

where s.e. $\left(\widehat{S}(t)\right)$ is the square root of the modified Greenwood's formula for the estimated variance.

We now plot the two modified Kaplan-Meier (K-M) survivor curves using S. This plot displays a difference in survival between the two groups. Figure 7.6 displays the estimated survival curves for the psychiatric inpatients. The psychiatric data is stored in a data frame named psychi. The S function Surv needs three arguments; it calls the left truncation entry, the observed time futime, and the censoring indicator status. This is very similar to the setup needed for an extended Cox model on page 193 in S.

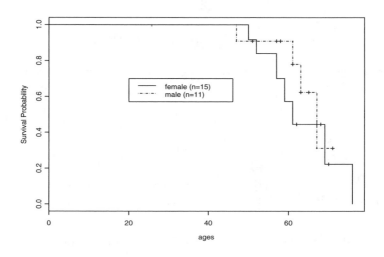

Figure 7.6 *The estimated survival curves for psychiatric inpatients under LTRC.*

```
> attach(psychi)
> psychi.fit0 <- survfit(Surv(entry,futime,status) ~ sex)
> plot(psychi.fit0,lwd=2,col=1,type="l",lty=c(1,3),
        xlab="ages",ylab="Survival Probability")
> legend(20,0.7,type="l",lty=c(1,3),
            c("female (n=15)","male (n=11)"),col=1)
> summary(psychi.fit0) # Displays the survival probability
                # table for each group. The output is omitted.
> psychi.fit0
        n events mean se(mean) median 0.95LCL 0.95UCL
sex=F 15    11 63.7    2.39     61      57      NA
sex=M 11     4 65.0    2.26     67      63      NA
> detach()
```

Results:

- Notice the estimated mean and median survival times of men are higher than those of females.
- The crossing modified K-M curves suggest that there is no significant difference in the lengths of lives between males and females. However, we notice that males have larger survival than females between early fifties and mid sixties. This suggests that the PH assumption is not satisfied.

To see the effect of mistreating LTRC data as RC data, we plot the two survival curves for each sex on the same graph.

```
> fem <- psychi[psychi[,1]=="F",] # Females only
> psychi.fit1 <- survfit(Surv(futime,status) ~ 1,data=fem)
> psychi.fit2 <- survfit(Surv(entry,futime,status) ~ 1,data=fem)
> male <- psychi[psychi[,1]=="M",] # Males only
> psychi.fit3 <- survfit(Surv(futime,status) ~ 1,data=male)
> psychi.fit4 <- survfit(Surv(entry,futime,status)~1,data=male)
> par(mfrow=c(1,2))
> plot(psychi.fit1,mark.time=F,conf.int=F,xlab="ages",
        ylab="Survival Probability",main="Female's Survival")
> lines(psychi.fit2,mark.time=F,lty=3)
> legend(locator(1),c("RC","LTRC"),lty=c(1,3))
> plot(psychi.fit3,mark.time=F,conf.int=F,xlab="ages",
        ylab="Survival Probability",main="Male's Survival")
> lines(psychi.fit4,mark.time=F,lty=3)
> legend(locator(1),c("RC","LTRC"),lty=c(1,3))
```

Figure 7.7 displays the estimated survival curves for each sex under the RC and LTRC data, respectively. It is clear that the estimated survival curve for female under the RC data setup lies above the corresponding estimated survival curve under the LTRC data setup. On the other hand, the two estimated survival curves for male coincide. WHY! This shows that mistreating the left truncation leads to overestimation of the survival probabilities.

7.3.2 Cox PH model for LTRC data

Suppose we have n iid observations $(X_i, Y_i, \delta_i, \underline{Z}_i)$, $i = 1, \ldots, n$, where X_i is the left truncation time, $Y_i = \min(T_i, C_i)$ the observed failure time, $\delta_i = 1$ if $X_i < T_i < Y_i$, or $\delta_i = 0$ if $X_i < Y_i < T_i$, and $\underline{Z}_i$ is the covariate. To estimate the regression coefficients with LTRC data, the partial likelihood for RC data (6.21) is modified to account for delayed entry into the risk set. It is easy to see that the **modified partial likelihood** for LTRC data in terms of all n observed times is given as follows:

$$L(\underline{\beta}) = \prod_{i=1}^{n} \left(\frac{\exp(\underline{z}_i'\underline{\beta})}{\sum_{l \in \mathcal{R}(y_i)} \exp(\underline{z}_l'\underline{\beta})} \right)^{\delta_i}, \tag{7.16}$$

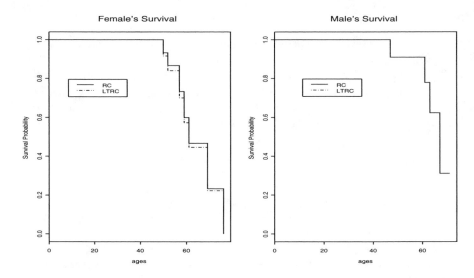

Figure 7.7　*The estimated survival curves under RC and LTRC data setups.*

where $\mathcal{R}(y_i)$ is the risk set at y_i defined in expression (7.10). With this modification, the methods discussed in Chapter 5 carry out. We shall illustrate the methods in the following psychiatric inpatients example.

```
> psychi.cox <- coxph(Surv(entry,futime,status) ~ sex)
> psychi.cox # Check if sex has a moderately large p-value.
      coef  exp(coef) se(coef)      z     p
sex -0.26      0.771      0.3 -0.866  0.39
```

The following output shows the three tests – LRT, Wald, and efficient score test – indicate there is no overall significant relationship between sex and survival time. That is, it does not explain a significant portion of the variation.

```
> summary(psychi.cox)
  Likelihood ratio test= 0.8    on 1 df,   p=0.372
  Wald test            = 0.75   on 1 df,   p=0.387
  Score (logrank) test = 0.77   on 1 df,   p=0.381
```

To see the effect of mistreating LTRC data as RC data, we run coxph by treating the psychi data as merely RC data and compare the result with what we obtained from LTRC.

```
> psychi.cox1 <- coxph(Surv(futime,status) ~ sex)
> psychi.cox1 # Check if sex has a moderately large p-value.
       coef  exp(coef) se(coef)      z     p
sex -0.222      0.801    0.299 -0.742  0.46
```

```
> summary(psychi.cox1)

Likelihood ratio test= 0.58  on 1 df,   p=0.445
Wald test            = 0.55  on 1 df,   p=0.458
Score (logrank) test = 0.56  on 1 df,   p=0.454

> detach()
```

The output above shows that the regression parameter estimate from RC setup, (-0.222), is not much different than that from LTRC setup, (-0.26). The estimated standard errors are almost identical (0.299 versus 0.3). The three tests – LRT, Wald, and efficient score test – also indicate there is no overall significant relationship between sex and survival time. That is, it does not explain a significant portion of the variation. In this small data set ($n = 26$), mistreating the LTRC data as RC data doesn't seem to have any significant differences in terms of the regression parameter estimate and its standard error estimate. However, in general it may lead to biased regression parameter estimates and their standard error estimates.

Result:

- As we noticed in Figure 7.6, the Cox PH model is not a good choice as the p-value is too big for each of the three tests.

Discussion

- Treating LTRC data as merely RC data may lead to a severe overestimation of the survival probability.

- Treating LTRC data as merely RC data may also lead to biased regression parameter estimates and their standard error estimates from Cox PH model.

- A key assumption for Cox PH model with LTRC data is that the left truncation X and the failure time T are independent, given the covariates $\underline{Z}$. If this assumption is violated, other methods are needed. See Keiding (1992) for more details.

- Depending on the interest of study, the left truncation variable can be used as a covariate. For example, if the time of interest had been the time to death since the admission to the Iowa hospital for psychiatric inpatients considered in Tsuang and Woolson (1977), one could have used the entry time as a covariate. Let's see what happens under this setup. We leave the details as Exercise 7.3.

- The psychiatric data we reanalyzed here is only a part of a much larger sample of 525 psychiatric inpatients in Tsuang and Woolson (1977). It seems that analyzing the original data would be worthwhile as the data set at hand is only one-twentieth of the original. The survival curves between males and females might be substantially different from what we have. Depending on the estimated survival curves, the Cox PH model may need to be considered again.

- The S function `survdiff` and `survReg` do not work for LTRC data. As far as we know, there is no commercial statistical software that performs log-rank test and parametric regression modelling for LTRC data. This needs to be done somewhere else.

- However, for the Cox PH model, the methods used for the RC data such as variable selection method by `stepAIC`, stratification, and diagnostic checking methods carry over well to LTRC data.

7.4 Exercises

A. *Applications*

7.1 Complete by hand the survivor estimate in expression (7.12) on page 207. Now use `survfit` to obtain the survivor estimate. Does your computation match the `survfit` output?

7.2 Complete by hand the variance estimate in expression (7.14) on page 207. Now use `survfit` to obtain the survivor estimate. Does your computation match the `survfit` output?

7.3 Run `coxph` assuming the entry time as a covariate. That is, run `coxph` with covariates, age at entry and sex. Start with two-way interaction model and find the best model. Draw a martingale residual plot. Is a cut point analysis necessary? If so, find a cut point and compare your results.

B. *Theory and WHY!*

7.4 This problem requires a closer look at expression (7.9).

   (a) Prove expression (7.9).
   (b) Find by hand the MLE of λ.
       Hint: To find the MLE by hand, take the derivative of the log of the likelihood function (7.9) with respect to λ and set it equal to zero. The solution is the MLE.
   (c) Find the MLE of λ by treating as if the data were RC data. How large is the difference between the two MLE's, one under LTRC data and the other under RC data?
       Hint: See page 71.

7.5 Answer the WHY! on page 209.
    Hint: See expression (7.10) and figure out the modified risk set.

CHAPTER 8

Censored Regression Quantiles
by Stephen Portnoy

An alternative to the Cox PH model

When the Cox PH model fails to hold, an alternative methodology called **censored regression quantiles** can find general departures from the PH model and important forms of heterogeneity. This chapter suggests censored regression quantiles as a useful complement to the usual Cox PH approach.

Objectives of this chapter:

After studying Chapter 8, the student should:

1. Understand the definition of conditional quantiles.

2. Become familiar with the regression quantile model and how to interpret it, especially in terms of population heterogeneity.

3. Know the basic properties of regression quantiles.

4. Understand the censored regression quantile model and how to use it to analyze censored survival data.

5. Know the difference between the censored regression quantile model and the Cox PH model. Especially, know what censored regression quantiles can provide for data interpretation that is lacking in an analysis based on a Cox PH model.

8.1 Introduction

The Cox PH model imposes structure on the hazard function. In a fundamental way, it considers **survival** as a stochastic process, and models the rate of change from one state to another. However, in many situations, both the basic data and the questions of interest involve survival times. Thus, it is often valuable to have flexible models that pertain directly to the survival times. In fact, one such approach is called the **accelerated failure time model** presented in Chapter 4.4 and posits a linear regression relationship between the logarithm of the survival times and the covariates. Emphasizing the fact that this is often more natural, Sir David Cox has written:

Of course, another issue is the physical or substantive basis for the proportional hazards model. I think that's one of its weaknesses, that accelerated life models are in many ways more appealing because of the quite direct physical interpretation the accelerated failure time model offers. See Reid (1994).

Although accelerated failure time models have been extended to allow for censoring, the major shortcoming with this approach has been that the Cox PH model permits flexible (nonparametric) estimation of the baseline hazard and, as a consequence, permits analysis of the survival function at any specific settings of the covariate values. Thus, it is necessary to extend the linear model approach to one that allows similar flexibility in analyzing conditional survival functions. Fortunately, the methodology of **regression quantiles** introduced by Koenker and Bassett (1978) provides just such flexibility. This methodology has recently been extended to allow for standard right censoring and, thus, to provide a flexible alternative to the Cox PH model. See Portnoy (2003).

8.2 What are regression quantiles?

Traditionally, statisticians carried out their analyses by taking the sample average as an estimate of the population mean. If some notion of variability was desired, one simply assumed normality and computed a standard deviation. Unfortunately, even after transformation, real data is rarely (if ever!) exactly normal, and it often displays some form of heterogeneity in the population. This was realized as early as 1890 by Sir Francis Galton, the inventor of the idea of regression. In his book, *Natural Inheritance* (1889, page 62) Galton chided those of his statistical colleagues who

> limit their inquiries to Averages and do not revel in more comprehensive views. Their souls seem as dull to the charm of variety as that of a native of one of our flat English counties, whose retrospect of Switzerland was that, if the mountains could be thrown into its lakes, two nuisances would be got rid of at once.

The question then is what can replace such mean analyses. In the same book, Galton argued that any complete analysis of the full variety of experience requires the entire distribution of a trait, not just a measure of its central tendency. Galton then introduced the empirical quantile function as a convenient graphical device for this purpose. That is, Galton suggested that heterogeneity among subpopulations could be identified and analyzed by considering the set of percentiles as a function of the probability, τ. The focus of this section is to show how the idea of univariate quantiles can be extended to regression settings, and especially to linear regression models.

8.2.1 Definition of regression quantiles

Formally, given a random variable Y of measurements for some population, the population quantile is defined to be the value $Q_Y(\tau)$ satisfying

$$P\{Y \le Q_Y(\tau)\} = \tau \quad \text{for } 0 \le \tau \le 1. \tag{8.1}$$

The cumulative distribution function (d.f.) is considered as more basic, and so the quantile function is typically defined as

$$Q_Y(\tau) = F_Y^{-1}(\tau) = \inf\{y : F_Y(y) \ge \tau\},$$

where $F_Y(y)$ is the (population) d.f.

Recall from Chapter 2 that the population quantile is estimated by a specific order statistic, or by a linear combination of two adjacent order statistics. For example, the sample median for a sample of even size is the average of the two middle order statistics. In fact, statisticians will generally allow any value between the two order statistics to be called a median. In this sense, the median is **not** unique. Note however that the length of the interval of nonuniqueness is just the distance between two successive observations, and in a sample of size n this distance is roughly proportional to $1/n$ (unless there is an interval about the median with zero probability). Since the statistical accuracy of the estimator of the median is proportional to $1/\sqrt{n}$ (just as for the mean), there is clearly no statistical implication of this nonuniqueness.

However, even if the quantile function provides a reasonable approach to the analysis of heterogeneity, the question remains of how to generalize the notion to more complicated settings like regression models. In terms of the population, there does seem to be a direct generalization in terms of conditional quantiles. If we observe a random response, Y, and wish to model the response as a (perhaps linear) function of a vector of explanatory variables, then one can take $\underline{x}$ to be the realization of a random vector $\underline{X}$ and consider the conditional distribution of Y given $\underline{X} = \underline{x}$ for each fixed $\underline{x}$. This permits us to define the conditional quantile of Y given $\underline{X} = \underline{x}$ as the inverse of the conditional d.f. Specifically, the conditional quantile, $Q_{Y|\underline{X}}(\tau; \underline{x})$, satisfies

$$P\{Y \le Q_{Y|\underline{X}}(\tau; \underline{x}) \mid \underline{X} = \underline{x}\} = \tau. \tag{8.2}$$

Traditional regression analysis posits a single regression curve, *e.g.*, the conditional mean function. Following Galton, we suggest that letting τ vary and considering the family of conditional quantile curves provides a much more complete picture of the data.

In general, the conditional quantile functions may not be linear in $\underline{x}$. This leads to difficulties in interpretation as well as nearly insurmountable computational issues if the dimension of $\underline{x}$ is large. Thus, it would be desirable to estimate conditional quantile curves under the assumption that they are linear in the $\underline{x}$-coordinates (at least, after some appropriate transformations). This will preserve the ease of interpretation offered by linear models; and, as will be shown, it will provide very fast computation. So we turn to the problem of how to estimate such linear regression functions, recognizing that the notion of quantile seems to require the perplexing task of ordering these regression lines.

For linear conditional quantiles, an especially useful solution to this problem was introduced by Koenker and Bassett (1978), who observed that a more fruitful approach to generalizing the concept of quantiles might employ an implicit characterization: the τth sample quantile may be expressed as the solution to the problem; choose ξ to minimize

$$R_\tau(\xi) = \sum_{i=1}^{n} \rho_\tau(Y_i - \xi), \tag{8.3}$$

where ρ_τ is the piecewise linear "check" function,

$$\rho_\tau(u) = u(\tau - I(u < 0)) = \tau\, u^+ + (1 - \tau)\, u^- . \tag{8.4}$$

Here u^+ and u^- are the positive and negative parts of u (taken positively).

One way to see that this is so is to note that $R_\tau(\xi)$ is a piecewise linear function and that the directional derivatives in both left and right directions from the order statistic corresponding to the τth-quantile must be nonnegative for this corresponding order statistic to be optimal. Some details of algorithms for minimizing (8.3) will be given in the next section. This use of the directional derivative (or what mathematicians call the subgradient conditions) will be exploited in a strong way in extending these ideas to censored observations in the next section.

Consider a general linear response model: let $\{Y_i, \underline{x}_i\}$ denote a sample of responses Y and explanatory variables $\underline{x}$ (in m dimensions) and suppose

$$Y_i = \underline{x}_i'\underline{\beta} + z_i, \quad i = 1, \ldots, n, \tag{8.5}$$

where $\underline{\beta}$ is an m-dimensional parameter and $\underline{z}_i$ is a random disturbance. As noted above, there is no natural notion of ordering the sample in terms of regression planes. However, the *implicit* approach to defining the sample quantiles via optimization does extend in a natural way, yielding the m-dimensional **regression quantiles**, $\underline{\hat{\beta}}(\tau)$, minimizing (over $\underline{\beta}$) the objective function

$$R_\tau(\underline{\beta}) \equiv \sum_{i=1}^{n} \rho_\tau(Y_i - \underline{x}_i'\underline{\beta}). \tag{8.6}$$

Following Koenker and Bassett (1978), extensive development has shown these regression quantiles to generalize successfully many of the fundamental properties of the sample quantiles. First note that these regression quantile parameters **depend** on the probability, τ. Specifically, the jth coordinate of $\underline{\hat{\beta}}(\tau)$ gives the marginal effect of a unit change in the jth explanatory variable, $x^{(j)}$, on the (conditional) τth-quantile of the response. If there is an intercept term in (8.5) and the errors are iid, then the intercept coefficient is a location shift of the quantile function for the error distribution, but the beta coefficients for the x's are **constant** in τ. Thus, if the β coefficients really change with τ, there must be heterogeneity in the population. This heterogeneity can be much more than simple heteroscedasticity (that is, unequal variances). It will represent the possibly varying effect of heterogeneity among subpopulations.

Specifically, the variation in the β's permits the structure of the linear relationship to change with the response quantile. For example, in a study relating student performance to the time spent on homework, it would not be surprising if the better performing students had a larger regression coefficient. Or in studies of animal or plant growth, it would not be surprising to find that those individuals whose growth is the fastest depend very differently on variables like nutrition or treatment than those growing more slowly. In fact, such heterogeneity seems even more likely to occur for responses that are time durations (see Koenker and Geling (2001) and several examples at the end of this chapter). As a consequence, it becomes especially important to extend the regression quantile approach to censored survival data.

It is important to consider the possible sources of heterogeneity in regression models. In most cases, such heterogeneity can arise from the presence of explanatory variables that are not included in the study. If such variables have an effect on the response and are correlated with the explanatory variables that are measured, then their effect must be observed as variation (over τ) of the β coefficients for those x's that were observed. Thus, regression quantiles provide a natural approach to analyzing as well as diagnosing population heterogeneity, especially if it is caused by inadequacies in the model.

8.2.2 A regression quantile example

To clarify these ideas, consider an example of an artificial data set based loosely on a study of how student performance depends on the amount of study time. Here the explanatory variable X is the average weekly number of hours spent on homework, and the response variable Y is the student's grade point average (GPA). For pedagogic purposes, a sample of size $n = 10000$ was simulated so that the conditional quantiles were in fact linear, but not parallel (as they would be if the errors were iid). First, consider the conditional quantiles at specific x-values corresponding to homework times of 5, 10, 15, 20, and 25 hours. In most data sets, there would not be enough observations at any one x-value to estimate an entire quantile function. But here (when rounded slightly), there were from 12 to nearly 150 observations at each of these values. A plot (for these x-values only) is given in Figure 8.1. The plot indicates sample quantiles at each x above corresponding to probabilities $\tau = .1, .25, .5, .75,$ and $.9$. These quantiles have been connected to produce piecewise linear conditional quantile functions. Note that the statistical accuracy of these estimates is only proportional to one over the square root of the number of observations in each sample. Some improvement in accuracy can be obtained by assuming the quantile functions are linear in x and fitting the 5 points for each quantile using least squares. The linear quantile estimates are given as dotted lines in the plot. Notice that though the data were simulated to have linear quantile functions, the discrepancy between the piecewise linear quantile functions and their linear approximations is noticeable. This is a consequence of the relatively small sample sizes at each of the 5 x-values. Now consider the entire data set. A plot is given in Figure 8.2.

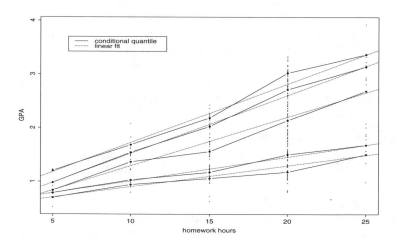

Figure 8.1 *Student performance vs. homework; conditional quantiles at x = 5, 10, 15, 20, 25. Curves (from bottom to top) correspond to quantiles τ = .1, .25, .5, .75, .9.*

The solid lines are the regression quantiles, $\hat{\beta}(\tau)$, for $\tau = .1, .25, .5, .75, .9$, computed by minimizing (8.6). Since $\sqrt{10000} \approx 100$, these lines are accurate to roughly 1% (or better). The dotted lines are the fitted lines from the previous plot. Notice the clear heterogeneity. Finally, Figure 8.3 shows one of the most important facets of regression quantile analysis: a plot of the regression coefficient $\hat{\beta}(\tau)$ against the probability τ. The plot was produced by computing the regression quantiles for $\tau = .01 : .99$, together with 95% confidence intervals based on asymptotic theory discussed in the next section, and then shading the region between the confidence bands. The fact that $\hat{\beta}(\tau)$ is so obviously nonconstant in τ demonstrates clear heterogeneity. Here, this heterogeneity is a form of "heteroscedasticity," or the variation in the standard deviation of the response with x. In fact, the data were simulated to have an error standard deviation proportional to x. One might interpret the results as follows: since the GPA's for the poorest students are essentially independent of study time, such students are apparently unable to make good use of study time. The better students seem much better able to make good use of study time, and their performance benefits in a way that is proportional to study

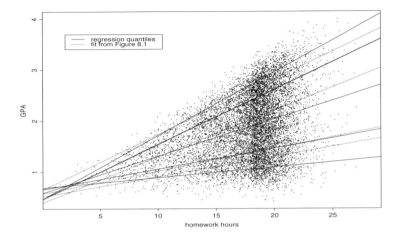

Figure 8.2 *Student performance vs. homework; regression quantiles. Curves (from bottom to top) correspond to quantiles* $\tau = .1, .25, .5, .75, .9$.

time. Notice that true conditional quantile functions must be monotonically increasing with τ. The fact that the regression quantile lines cross in the lower

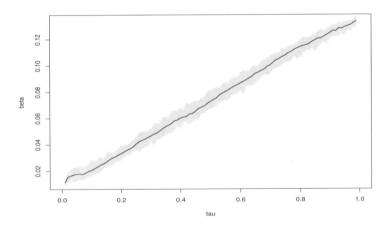

Figure 8.3 *Student performance vs. homework; the regression quantile homework coefficient as a function of* τ.

left corner of Figure 8.2 is an artifact of random variation in the statistical estimates. There are clearly very few of these extremely poor students, and thus the regression quantiles are not sufficiently well estimated in this region to avoid crossing. Note that the dotted lines in Figure 8.2 cross for much larger x-values – thus indicating their relatively greater statistical variability.

Also note that the crossing problem can often be alleviated by choosing better models. If we assume that Y is a (3-parameter) quadratic function of x, then the quantile functions do not cross at all on the domain of the observed data, even though the quadratic coefficients are very small (so that the quadratic curves remain very nearly linear; see Figure 8.4).

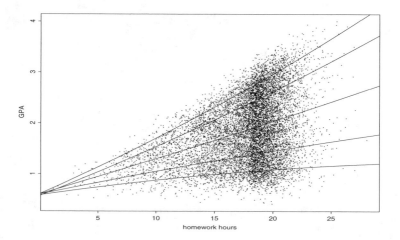

Figure 8.4 *Student performance vs. homework; regression quantiles for the quadratic model:* Y *(GPA)* $= \beta_0 + \beta_1 x + \beta_2 x^2$ *where* x = *homework hours.*

8.3 Doing regression quantile

The algorithm and asymptotic inference

Almost all of the properties of the regression quantiles arise from one fundamental fact: although the objective function (8.6) appears to be quite complicated, each regression quantile solution is exactly the solution of a linear programming problem. In mathematical optimization, a linear programming problem is simply to maximize (or minimize) a linear combination of variables, subject to a set of linear inequalities.

To see that a regression quantile solution also solves a linear programming problem, introduce auxiliary variables $\{v_i\}$ and $\{w_i\}$ for $i = 1, \ldots, n$ satisfying

$$v_i - w_i = Y_i - \underline{x}_i'\underline{\beta}, \quad \text{where} \quad v_i \geq 0, \quad w_i \geq 0. \tag{8.7}$$

Now express the objective function (8.6) by writing

$$\rho_\tau(Y_i - \underline{x}_i'\underline{\beta}) = \rho_\tau(v_i - w_i) = (v_i - w_i)\left(\tau - I(v_i - w_i < 0)\right).$$

It is now clear that any value of $(v_i - w_i)$ can be achieved by taking either

$v_i = 0$ or $w_i = 0$ (or both) **without** increasing the objective function. If we take one of v_i or w_i to be zero, then from the equality $Y_i - \underline{x}_i' \underline{\beta} = v_i - w_i$, the other must become either the positive or negative part of $Y_i - \underline{x}_i' \underline{\beta}$. Thus, at a solution, the objective function becomes

$$\sum_{i=1}^{n} \left(\tau v_i + (1 - \tau) w_i \right) . \qquad (8.8)$$

WHY! Therefore, the regression quantile solution minimizes a linear combination of the $(2n + m)$ variables $(\{v_i\}, \{w_i\}, \underline{\beta})$ subject to the linear constraints (8.7).

To see what this means, consider a linear programming problem in two dimensions: minimize $ax + by$ subject to a family of linear constraints. The constraints require (x, y) to lie in the intersection of a family of half-spaces; that is, in a **simplex** S as indicated in Figure 8.5. The sets $\{(x, y) : ax + by = c\}$ form a family of parallel straight lines (as c varies); and so to minimize $ax+by$, it suffices to move the line $ax + by = c$ parallel to minimize c. That is, in Figure 8.5, choose the point of S that just touches the line $ax+by = c$ closest to the origin. Generally, this point will be a vertex of S (unless $ax + by = c$ is parallel to one of the boundary edges of S, at which the entire boundary segment will be solutions). A **vertex** of S is defined by taking two of

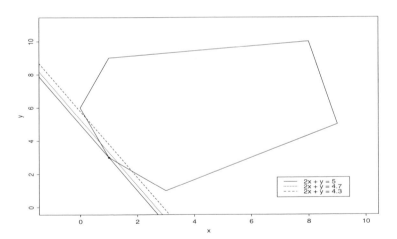

Figure 8.5 *Linear programming simplex of points satisfying linear constraints.*

the inequality constraints to be equalities. In terms of the regression quantile problem in m dimensions, this means that $v_i = w_i = 0$ for exactly m indices. Equivalently, a regression quantile solution must have m zero residuals; that is, it must fit m observations exactly. In fact, there will be exactly m zero residuals unless more than m observations lie on the same m-dimensional plane, which happens with zero probability if the response has a continuous distribution. The specific set of m observations is defined by the minimization problem. To find the set, consider the classical **simplex** algorithm for linear programming problems. Start at an arbitrary vertex, and then move in the direction of steepest descent (in Figure 8.5, to the southwest) from vertex to vertex until reaching the optimal solution. A move from one vertex to the next is called a **pivot** and consists of replacing one of the constraints defining the first vertex by another one to define the next vertex. In the language of regression quantiles for an m-dimensional parameter, one simply replaces one of the m observations fit perfectly (that is, with zero residual) by another observation that is to have zero residual at the next vertex.

Suppose we have found a solution at $\tau = \tau_0$, and now consider what happens as the parameter τ changes. Mathematically, in the objective line, $ax+by = c$, the coefficients a and b now depend linearly on τ. Thus, as τ changes, the slope of the line changes continuously; that is, the line tilts. The optimal vertex at $\tau = \tau_0$ will remain optimal for an interval of τ values until the line tilts (with this optimal vertex held fixed) to lie along the adjoining boundary segment. At this point, both vertices (and all points along the boundary segment) will be optimal solutions. Furthermore, if τ continues to change, the line continues to tilt, and the next vertex will become a unique solution for another interval of τ values. Thus, the regression quantile solutions form a piecewise constant function of τ, with breakpoints at the τ values for which two adjacent vertices are both solutions. Note that this nonuniqueness is really no more problematic than for one-sample quantiles. It is in fact a direct generalization of the fact that the median of an even number of observations is any value between the two middle observations.

Thus, to find the regression quantile solutions, the major problem is to identify the interval over which a solution is optimal. This is a direct consequence of the **subgradient conditions**. Suppose we are at a specific solution, $\hat{\beta}(\tau_0)$. Simply take directional derivatives of the objective function (8.6) with respect to each coordinate of β (that is, in the positive and negative direction). The condition that these directional derivatives must be nonnegative at a solution gives $2m$ inequalities that are **linear** in τ. These inequalities provide m upper bounds and m lower bounds for the values of τ for which $\hat{\beta}(\tau_0)$ remains optimal. Thus, the largest of the m lower bounds and the smallest of the m upper bounds provide the interval around τ_0 where $\hat{\beta}(\tau_0)$ remains optimal. To find the next solution (say for the next larger τ value), simply delete the observation corresponding to the inequality giving the smallest upper bound (that is, the first inequality constraint on τ to be violated), and replace it with the "nearest" observation.

As a consequence, to find the entire set of all regression quantile solutions, one begins by using the simplex algorithm to solve for the initial quantile, $\hat{\beta}(0)$ (at $\tau = 0$); and then successively applies the above pivoting method to find successive breakpoints $\{\hat{\tau}_j\}$, and the corresponding solutions. The computation of the subgradient inequalities and the pivoting process will be illustrated in the next section in the context of censored regression quantile. Some exercises are given to lead the student through this process.

In Section 8.5, the censored regression quantiles will be defined as "weighted" regression quantiles with the weights $\{w_i \geq 0\}$ determined empirically. Note that the weighted version of the objective function (8.6) is just

$$R_\tau^w(\underline{\beta}) = \sum_{i=1}^{n} w_i \, \rho_\tau(Y_i - \underline{x}_i'\underline{\beta}) = \sum_{i=1}^{n} \rho_\tau(w_iY_i - w_i\underline{x}_i'\underline{\beta}) , \qquad (8.9)$$

and so weighted regression quantiles can be computed by applying the unweighted algorithm to the responses $\{w_iY_i\}$ and design vectors $\{w_i\underline{x}_i\}$.

Many basic properties now follow from these considerations. It can be shown that the number of observations below the τth regression quantile plane, $y = \underline{x}'\hat{\beta}(\tau)$, is within m of the appropriate number, τn. That is, the empirical regression quantile plane lies above a fraction τ of the data with accuracy m/n. Note that this accuracy is much better than the usual statistical accuracy of $1/\sqrt{n}$. One can also show that the regression quantile $\underline{\bar{x}}'\hat{\beta}(\tau)$ is in fact monotonically nondecreasing in τ. That is, there can be no crossing at (or near) the mean, $\underline{\bar{x}}$, of the explanatory variables. The subgradient condition also implies that $\hat{\beta}(\tau)$ depends on the responses $\{Y_i\}$ **only** through the sign of the residual. Hence, if the responses are perturbed so that the sign of the residuals does not change, then the regression quantiles remain unchanged. See Exercise 8.6(b).

One other crucial consequence of the subgradient conditions is the following: in random regression problems, the number of breakpoints is of order $n \log n$ with probability tending to one. See Portnoy (1991a). In practice, one rarely needs more than $3n$ breakpoints unless n is much larger than 10^4. Thus, the computation of the entire regression quantile process requires only a constant times n pivots, and is quite quick.

Finally, the subgradient conditions permit the development of an asymptotic theory for regression quantiles. Specifically, under relatively mild conditions, one can show that $\hat{\beta}(\tau)$ has an asymptotic normal approximation, and that furthermore, this approximation is uniform in $\epsilon \leq \tau \leq 1 - \epsilon$ for any fixed $\epsilon > 0$. Specifically, we have the following limiting distribution result: if the conditional quantiles are in fact linear, and other regularity conditions hold, then (conditionally on $\{\underline{x}_i\}$),

$$n^{\frac{1}{2}} \left(\hat{\underline{\beta}}(\tau) - \underline{\beta}(\tau) \right) \stackrel{a}{\sim} \mathcal{N}\left(0, \, D_\tau^{-1}\Sigma D_\tau^{-1} \right) , \qquad (8.10)$$

where

$$D_\tau = n^{-1} \sum_{i=1}^{n} \underline{x}_i\underline{x}_i' f_i\left(F_i^{-1}(\tau) \right)$$

with F_i denoting the d.f. for the ith observation, and f_i its density; and where

$$\Sigma = \tau(1 - \tau)\,\frac{1}{n}(X'X) = \tau(1 - \tau)\,\frac{1}{n}\sum_{i=1}^{n} \underline{x}_i \underline{x}_i'\,.$$

Recall that when the pair $(\underline{X}, Y)$ is assumed to have a joint distribution (in $\Re^{m+1}$), then F_i and f_i are the conditional d.f. and density of Y given $\underline{X} = \underline{x}_i$, respectively. Thus, the density of the response at the specified conditional quantile plays a crucial role in the asymptotics. Specifically, this requires that the response distribution not be discrete.

These asymptotics can be used to develop appropriate inferential methodologies. Specifically, hypothesis tests and confidence intervals have been developed and shown to be generally good approximations in numerous examples and simulation experiments. For some recent examples of this asymptotic development, see Portnoy (1991b), Gutenbrunner $et\ al.$ (1993), Koenker (1994) and Koenker and Machado (1999). It also justifies the bootstrap method for finding confidence intervals described in Section 8.6.

Remark:

The past 20 years has seen major breakthroughs in computational methods for very large linear programming problems. These approaches are called interior point methods, as they approximate the constrained linear problem by a smooth problem that permits one to approach the vertex from within the simplex region in a very fast and accurate manner. For the regression quantile problem, these methods can be combined with a stochastic preprocessing step to provide an algorithm that is provably faster than least squares methods (with probability tending to one) when the sample size is very large and the number of parameters is moderate. In fact, the algorithms developed in Portnoy and Koenker (1997) provide computation times comparable to those required by modern least squares methods for all sample sizes investigated, and strictly faster in some cases with $n > 10^5$. In this sense, quantile regression can be considered as even simpler than least squares.

8.4 Censored regression quantile model and Cox model

We extend the regression quantile approach to the case of right-censored data. To describe the basic models, consider the response variables to be the survival times $\{T_i\ :\ i = 1, \cdots, n\}$ that would generally be censored and that depend on m-dimensional covariates $\{\underline{x}_i \in \Re^m\ :\ i = 1, \cdots, n\}$. It is natural to consider the pairs $\{(T_i, \underline{x}_i)\}$ as a multivariate iid sample in $\Re^{m+1}$; but since our approach is based on conditional quantiles, we will develop the theory conditionally on $\{\underline{x}_i\}$. That is, $\{\underline{x}_i\}$ will be taken to be fixed; and when needed, $P_{\underline{x}_i}$ will denote the conditional probability as a function of $\underline{x}_i$. In applications,

T_i could be some function of the survival times: the "log" transformation providing the usual accelerated failure time model. See Chapter 4.4.

As discussed in Section 8.2, the basic regression quantile model specifies that the conditional quantiles of T_i given $\underline{x}_i$ obey a linear regression model. Specifically, assume there are real-valued functions $\{\beta(\tau)\}$ for $0 \le \tau \le 1$ such that $\underline{x}'_i\beta(\tau)$ gives the τth conditional quantile of T_i given $\underline{x}_i$; that is,

$$P_{\underline{x}_i}\left\{T_i \le \underline{x}'_i\beta(\tau)\right\} = \tau, \quad i = 1, \ldots, n. \tag{8.11}$$

Here, f_i and F_i will denote the conditional density and d.f. for T_i given $\underline{x}_i$, respectively. Asymptotic theory will require certain smoothness conditions on the densities and regularity on the design vectors. Note that the family densities for $\{T_i\}$ need **not** be a location-scale family. These distributions may be quite different for different observations. All that is needed is for the conditional quantiles to be linear in the parameters in (8.11).

In addition, there are censoring times, $\{C_i : i = 1, \cdots, n\}$. The distribution of C_i may depend on $\underline{x}_i$, but conditionally on $\underline{x}_i$, T_i and C_i are independent. This gives the general random censoring model. Define the censored (observed) random variables and censoring indicator by

$$Y_i = \min\{T_i, C_i\} \quad \text{and} \quad \delta_i = I_{\{T_i \le C_i\}}, \tag{8.12}$$

respectively. Again, the triples $\{(T_i, C_i, \delta_i)\}$ are often considered as a random sample. Note that although the conditional quantiles for $\{T_i\}$ are assumed to be linear (in β), the quantiles for $\{Y_i\}$ need not satisfy any linear model (nor need have any special form). As usual, all inference on $\beta(\tau)$ must be expressed in terms of the observable censored values $\{Y_i\}$, even though the parameters $\beta(\tau)$ pertain to the model for the (latent) survival times $\{T_i\}$.

Traditionally, the model used for analysis of censored survival times is the Cox PH model on the hazard function:

$$h_i(t) \equiv h(t|\underline{x} = \underline{x}_i) = -\frac{\mathrm{d}}{\mathrm{d}t} \log S_i(t) = \frac{f_i(t)}{1 - F_i(t)}. \tag{8.13}$$

Here $F_i(t)$ and $f_i(t)$ denote the conditional d.f. and density of T_i given $\underline{x}_i$, and $S_i(t)$ is the conditional survival function given $\{\underline{x}_i\}$; i.e., $S_i(t) = 1 - F_i(t)$.

Recall from Chapter 4.3, expression (4.7), that the Cox PH model takes $\log(h_i)$ to be additively separable in t and linear in the covariates. Specifically, $h_i(t)$ is modelled as

$$h_i(t) = h_0(t)\, e^{\underline{x}'_i\beta}, \quad i = 1, \ldots, n, \tag{8.14}$$

where $h_0(t)$ is the baseline hazard function. In this model, the conditional quantiles have a rather special form: by integrating,

$$S_i(t) = \exp\left\{-H_0(t)e^{\underline{x}'_i\beta}\right\}, \quad \text{where} \quad H_0(t) \equiv \int_0^t h_0(s)\, ds.$$

So the conditional quantile for Y at $\underline{x}$ becomes

$$Q_{Cox}(\tau \mid \underline{x}) = H_0^{-1}\left(-\log(1-\tau)\,e^{-\underline{x}'\underline{\beta}}\right) \ . \qquad (8.15)$$

Note that the exponential form of the Cox PH model requires that the quantiles (for $0 \leq \tau \leq 1$) all have a special form depending on H_0: In particular, they are all specific monotone functions of H_0. As noted in Section 8.1, this greatly restricts the behavior of the quantile effects. Section 8.6 provides examples of heterogeneity that violate the Cox PH model. The ADDICTS data of Chapter 7.1 also shows heterogeneity as described in Exercise 8.2.

One other important difference between the two approaches is the following: the fact that the regression quantile analysis fits all quantiles necessarily reduces the statistical accuracy of the estimates (especially for larger quantiles, where more of the censored observations play a role). Thus, effects are sometimes less significant (especially for larger τ-values) than suggested by the Cox method. In other words, the fact that the Cox effects are global (that is, the coefficients are constant) tends to provide a more powerful analysis if the model is correct. However, this comes at the cost of possibly missing important sources of heterogeneity: if the Cox PH model is wrong, the analysis may have substantial biases. On the other hand, the regression quantile approach is most valuable exactly when such heterogeneities are present. When the regression quantile model fits better than the Cox PH model, effects will tend to be more significant for the regression quantile approach as indicated in Section 8.6.3.

One disadvantage of the quantile approach appears when the covariates can be time-dependent. The instantaneous nature of the hazard function allows the extended Cox model to apply. However, it is not possible to identify the conditional quantile function for the survival times if the covariates change with time. The fundamental reason is that for statistical purposes it is not possible to condition on future (unknown) values of the covariate. Some progress may be possible in special cases of time-varying covariates. For example, if measurements are only taken at a few discrete times, then it might be reasonable to assume that the covariates (and, hence, the conditional quantile models) remain fixed between the measurements. Then the survival times between the measurements could obey a fixed regression quantile model, and the total survival time could be analyzed as a sum of regression quantile models. However, such approaches remain to be developed.

The main point here is that the quantile model and the Cox PH model are *complementary*. If the survival times are natural units of analysis, then the quantile model (which posits linearity for survival times) is likely to provide better results (as suggested by the quote of Cox at the beginning of this chapter). But in many cases (especially those with time-dependent covariates), the mortality process is the natural unit of analysis and the Cox PH model (which posits log-linearity of the hazard function) would then seem more reasonable. In fact, it is easy to apply both models in most problems; and the combination would be significantly more informative than restricting to any one model. Note that the two models are really different. If we are sure of the model, then one is right and the other wrong. However, we rarely believe that modelled relationships are exactly linear. Thus the use of both approaches (and the consequent variation in the point-of-view) provides not only a broader perspective, but is much less likely to miss important relationships.

8.5 Computation of censored regression quantiles

The algorithm here generalizes the Kaplan-Meier (K-M) estimator of Chapter 2 directly to the regression quantile setting. It reduces exactly to the K-M estimator when there is a single sample; and when there is no censoring, it is closely related to the parametric programming method described in Koenker and d'Orey (1987). It depends on a combination of two crucial ingredients: (1) a new way of viewing the **redistribute-to-the-right** form of the K-M estimator as a recursively reweighted empirical survival function, and (2) a fundamental calculation showing that the weighted subgradient inequalities remain piecewise linear in the quantile, τ, and thus permitting the development of an effective version of simplex pivoting.

8.5.1 The new Kaplan-Meier weights

As shown in Chapter 2.1, one way to compute the K-M estimator is to redistribute the mass of each censored observation to the observations on the right and then to compute the (recursively reweighted) empirical d.f. Basically, one redistributes the mass at C_i to observed values above C_i. The fundamental insight here is that since the empirical quantile function for any τ depends only on the signs of the residuals, the mass at C_i may be redistributed to any point above all the data (*e.g.*, infinity), and not necessarily to specific observations above C_i. The advantage of this insight is that the reweighting then applies in more general regression settings where it is not known which observations lie "above C_i."

The basic problem now is to estimate the censoring probability, $P\{T_i > C_i\}$, for each censored observation. From a regression quantile perspective, this means finding the value $\hat{\tau}_i$ at which the quantile function, $\underline{x}'_i\beta(\tau)$, crosses C_i. For each censored observation ($\delta_i = 0$), if $\hat{\tau}_i$ is given, then weights $w_i(\tau)$ can be defined for $\tau > \hat{\tau}_i$ as follows:

$$w_i(\tau) = \frac{\tau - \hat{\tau}_i}{(1 - \hat{\tau}_i)}. \tag{8.16}$$

Note that this is just the conditional probability that the observation lies below the τth-quantile **given** that it lies above C_i.

These weights can be used to define a weighted regression quantile problem (8.9) with weighted "pseudo-observations" at C_i (with weight $w_i(\tau)$) and at $+\infty$ (with weight $(1 - w_i(\tau))$). If τ_0 is fixed and all $\hat{\tau}_i < \tau_0$ are defined, this weighted regression quantile problem can be solved for $\tau > \tau_0$. This permits the recursive definition of $\hat{\tau}_i$ for the next censored observation that is crossed. Since the recursive definition of the weights (8.16) depends only on the breakpoints of the empirical quantile function, it suffices to describe how they are computed for the classical K-M estimator, and this will be done in the following section.

8.5.2 The single-sample algorithm

Consider a single sample with $\{Y_i = \min(T_i, C_i), \ \delta_i\}$ observed, but with no covariates. Temporarily think of the K-M technique as estimating the quantile function instead of the survival function. This is just a matter of associating the time points with one minus the probability jumps in the output of the K-M algorithm. Specifically, consider the ordered observations, $Y_{(1)} < Y_{(2)} < \ldots < Y_{(n)}$. If the smallest $j - 1$ observations $(Y_{(1)}, Y_{(2)}, \ldots, Y_{(j-1)})$ are all uncensored, then for $\tau \leq (j - 1)/n$ the K-M empirical quantile is just the inverse of the usual empirical d.f., and so has an increase of size $Y_{(i)} - Y_{(i-1)}$ at probability values, i/n for $i = 1, \ldots, (j - 1)$. Now, consider the smallest censored observation, say $Y_{(j)} = C_{(j)}$. How can we define the jump in the empirical quantile function for $\tau > (j - 1)/n$, i.e., for subsequent uncensored observations? (Note: the K-M estimator does **not** associate any positive probability with censored observations, and so all jumps in the quantile function must correspond to uncensored observations). At the first censored point, let $\hat{\tau}_1$ denote the accumulated probability up to the first censored observation: $\hat{\tau}_1 = (j - 1)/n$. Now consider splitting the censored observation into two observations with weights depending on the subsequent value of τ: one at the censored observation, $C_{(j)}$, with weight $w_1(\tau) = (\tau - \hat{\tau}_1)/(1 - \hat{\tau}_1)$ and the other at $+\infty$ with weight $1 - w_1(\tau) = (1 - \tau)/(1 - \hat{\tau}_1)$. Note that $w_1(\tau)$ estimates the conditional probability that $T_i \leq \hat{F}^{-1}(\tau)$ given $T_i > C_i$. This process can now be iterated recursively: at each τ, weight earlier (crossed) censored observations by $w_\ell(\tau)$, give weight $1 - w_\ell(\tau)$ to $+\infty$ (for the ℓ-th crossed censored observations), and give unit weight to all other observations (both censored and uncrossed uncensored observations). Since the empirical quantile function depends only on the sign of the residuals, it follows (with a bit of calculation) that this process will reproduce the inverse of the d.f. corresponding to the K-M estimator (just as for the classical "redistribute-to-the-right" algorithm). The K-M estimator of the survival function is then just 1 minus this d.f. estimator. The essential point is the simple observation that since the quantile estimate depends **only** on the sign of the residual, the mass to be redistributed may be assigned to any sufficiently large value, say $+\infty$.

As an example, consider estimating the survival function based on the single sample with 10 observations at $y = 1, 2, ..., 10$. Suppose the observations $y = 5$, $y = 7$, and $y = 8$ are censored (that is, known only to lie above the respective values: 5, 7, and 8). The survival estimator based on the weighting scheme above is developed as follows: Since the first four observations are uncensored, there is a jump of size .1 at each of these four observations. Thus, the survival estimator is $(1 - .1) = .9$, $(1 - .2) = .8$, $(1 - .3) = .7$, and $(1 - .4) = .6$ at these four points. Notice that each observation is the τth-quantile for an interval of τ-values. For example, $y = 4$ is the τth-quantile for $.3 \leq \tau \leq .4$.

The next observation, $y_5 = 5$, is censored and so we define the weighting constant by (8.16) using the breakpoint $\hat{\tau}_{(4)} = .4$ to define $\hat{\tau}_1 = .4$ for the first censored observation. We then split the censored observation $y_5 = 5$ into an observation at 5 with weight $(\tau - .4)/(1 - .4)$ and one at $+\infty$ with weight $(1 - \tau)/(1 - .4)$. There is no jump at $y = 5$, so the survival function jumps to the value $(1 - \tau)$ $(= (1 - \hat{\tau}_{(6)}))$ at $y = 6$. To find $\hat{\tau}_{(6)}$, we need to find the interval on which the observation $y = 6$ is the τth-quantile. To do this, consider the objective function (8.3), and consider the directional derivatives as y decreases from $y = 6$ and as y increases from $y = 6$. See Figure 8.6.

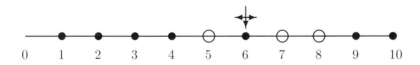

Figure 8.6 *Contributions to the directional derivatives.*

As y decreases from $y = 6$, there are four smaller (uncensored) observations (each weighted fully) and one censored observation that was split with weights $(\tau - .4)/.6$ at $y = 5$ and weight $(1 - \tau)/.6$ at $+\infty$. Note that the left directional derivative of ρ_τ is $-(1 - \tau)$, since as y decreases we are getting closer to observations with negative residual, each of which contributes the slope $(1 - \tau)$. Thus, the contribution of these smaller observations to the directional derivative of the objective function is

$$-\left(4 + \frac{\tau - .4}{.6}\right)(1 - \tau).$$

Since we are moving y to the left, the observation $y = 6$ is counted with all observations above $y = 6$. These are all fully weighted (since even the censored observations must be above $y = 6$). So, similarly, the contribution to the directional derivative from larger values (**including** the value at $+\infty$) is

$$\left(5 + \frac{1 - \tau}{.6}\right)\tau.$$

Thus, the condition that the left directional derivative must be positive for

$y = 6$ to be optimal requires

$$\left(5 + \frac{1 - \tau}{.6}\right)\tau - \left(4 + \frac{\tau - .4}{.6}\right)(1 - \tau) \geq 0.$$

Analogously, calculating the directional derivative to the right yields

$$-\left(4 + \frac{1 - \tau}{.6}\right)\tau + \left(5 + \frac{\tau - .4}{.6}\right)(1 - \tau) \geq 0.$$

Solving these two inequalities gives $\tau \geq .4$ and $\tau \leq .52$. Thus, $\hat{\tau}_{(6)} = .52$, and the survival function jumps down to $(1 - .52) = .48$ at $y = 6$. Note that in solving the inequalities, the $\tau(1 - \tau)$ terms cancel, thus leaving equations that are **linear** in τ. The fact that these are now linear equations (and **not** quadratic) is crucial for permitting linear programming methods to be applied in more general regression problems.

The next two observations are censored and so are assigned initial weights $\hat{\tau}_2 = \hat{\tau}_3 = .52$ for the second and third censored observations (for $\tau > .52$). As above, the next breakpoint of the survival function occurs at $y = 9$ (the next uncensored point) and satisfies

$$\left(2 + \frac{1 - \tau}{.6} + \frac{2(1 - \tau)}{.48}\right)\tau - \left(5 + \frac{\tau - .4}{.6} + \frac{2(\tau - .52)}{.48}\right)(1 - \tau) \geq 0$$

$$-\left(1 + \frac{1 - \tau}{.6} + \frac{2(1 - \tau)}{.48}\right)\tau + \left(6 + \frac{\tau - .4}{.6} + \frac{2(\tau - .52)}{.48}\right)(1 - \tau) \geq 0,$$

which gives $.52 \leq \tau \leq .76$. Thus, the survival function jumps to $(1 - .76) = .24$ at $y = 9$. Finally, it jumps to zero at $y = 10$. This is exactly the K-M estimate.

8.5.3 The general censored regression quantile algorithm

The above approach to the K-M estimator generalizes directly to the regression quantile model. The specific steps of the censored regression quantile computation are as follows:

1. Choose the first breakpoint, $\hat{\tau}_1$, and compute the initial quantile function $\hat{\beta}(\tau)$ for $0 \leq \tau \leq \hat{\tau}_1$ by the usual (uncensored) regression quantile algorithm (Koenker and d'Orey, 1987). This assumes that none of the basis elements defining this first solution are censored. Clearly such censored observations only imply that the corresponding survival time is above the bottom of the data. Any such censored values provide no information and can be deleted (as is done by the K-M estimator).

2. Let J be fixed (less than the number of censored observations), and suppose we have computed the breakpoints $\{\hat{\tau}_j : j \leq J\}$ for the "smallest" J censored observations. Suppose also that we have computed the regression quantiles $\hat{\beta}(\tau)$ for $\tau \leq \hat{\tau}_J$, and let J be the ℓ-th breakpoint ($\hat{\tau}_J = \hat{\tau}_{(\ell)}$). To

compute the next breakpoint (or breakpoints), and the regression quantile function, $\hat{\beta}(\tau)$ for $\hat{\tau}_J \leq \tau \leq \hat{\tau}_{J+1}$, simply carry out weighted simplex pivots until the next censored observation is crossed. To describe this more precisely, let M denote the indices for censored observations that have previously been crossed. That is, each $i \in M$ corresponds to the $j(i)$th censored observation encountered by the algorithm; and for each $i \in M$, a value $\hat{\tau}_j \equiv \hat{\tau}_{j(i)}$ has been assigned to each of these censored observations to provide the weight $w_{j(i)}(\tau)$ given by (8.16). Then, for $\tau > \hat{\tau}_J$ but such that $\hat{\beta}(\tau)$ crosses no other censored point not already in M, $\hat{\beta}(\tau)$ is chosen to minimize the weighted regression quantile objective function,

$$
\sum_{i \notin M} \rho_\tau(Y_i - \underline{x}_i'\underline{\beta}) \quad + \quad \sum_{i \in M} \left\{ w_{j(i)}(\tau) \rho_\tau(C_i - \underline{x}_i'\underline{\beta}) \right.
$$
$$
\left. + \quad (1 - w_{j(i)}(\tau)) \rho_\tau(Y^* - \underline{x}_i'\underline{\beta}) \right\}, \qquad (8.17)
$$

where Y^* may be taken to be $+\infty$ or any value sufficiently large to exceed all $\{\underline{x}_i'\underline{\beta} : i \in M\}$, and ρ_τ is the quantile loss (8.4). Computing directional derivatives with respect to $\underline{\beta}_k$ in the positive and negative directions gives m gradient conditions of the form $c_1 \leq a + b\tau \leq c_2$ (with b positive). The details of this computation are given in Portnoy (2003); and this is where the linearity in τ is crucial. Solving the subgradient conditions determines $\hat{\tau}_{(\ell+1)}$ (as in the one-sample example above), and a single simplex pivot defines $\hat{\beta}(\tau)$ for $\hat{\tau}_{(\ell)} \leq \tau \leq \hat{\tau}_{(\ell+1)}$. Continuing these subgradient calculations and pivots determines $\hat{\beta}(\tau)$ until the next previously unweighted censored observation is crossed; that is, until there is a new censoring point, C_{i^*}, with $C_{i^*} < \underline{x}_{i^*}'\hat{\beta}(\tau_{(\ell^*+1)})$, where i^* is the index of the next censored observation that is crossed, and ℓ^* is the number of the last pivot before C_{i^*}. We now define $j(i^*) = J + 1$ and $\hat{\tau}_{J+1} = \tau_{(\ell^*)}$. Note that previously crossed censored observations have a positive weight associated with the censoring point, C_i (for $j(i) \leq J$), and so are treated as regular known observations (though with weights less than 1), for $\tau \geq \hat{\tau}_J$.

3. Now, note that all censored observations that were not reweighted (and split) before $\hat{\tau}_J$ must have had positive residuals $(C_i > \underline{x}_i'\hat{\beta}(\hat{\tau}_J))$. At this point, as noted above, at least one (and perhaps several) of these censored observations are crossed (so that $C_{i^*} < \underline{x}_{i^*}'\hat{\beta}(\hat{\tau}_{J+1})$ for at least one and perhaps more than one value i^*). These observations must now be split and reweighted. We define the corresponding $\hat{\tau}_j \equiv \tau^*$, $j = J + 1, \ldots$ for all of the observations crossed at this pivot, where $\tau^* = \tau_{(\ell^*)}$ is the τ-breakpoint corresponding to the last pivot before any new censored observations were crossed. These $\hat{\tau}_j$ for the newly crossed censored points are now used to define future weights using (8.16). Lastly, we split each of the newly crossed censored observations by giving weight $w_i(\tau)$ to the newly crossed censored observation, C_i, and adding weight $1 - w_i(\tau)$ to $+\infty$. We can now continue pivoting exactly as in step 2.

4. The algorithm stops when either the next breakpoint equals 1, or when only

nonreweighted censored observations remain above the current solution. In this latter case, the last breakpoint, $\hat{\tau}_K$, will be strictly less than unity (in fact, it will be one minus the sum of all the weights redistributed to infinity). This is exactly analogous to the K-M estimate when the largest observations are all censored, and the survival distribution is defective.

This approach requires approximately a constant times $n \log n$ pivots, essentially as many as for the uncensored parametric programming calculation of the entire regression quantile process (Portnoy, 1991a). Thus, it is extremely quick for modest samples, and should be entirely adequate for sample sizes up to several thousand or so. See Section 8.6. There are several technical complications that are addressed in detail in Portnoy (2003). In addition, the current software permits the specification of a grid of τ-values (say $\tau = .01 : .99$) so that the censored regression quantiles are computed only along this grid. This approach is somewhat quicker in large data sets (of size greater than 1000).

8.6 Examples of censored regression quantile

After a brief description of the current software, two examples will be presented. The first is the CNS lymphoma data set, which was analyzed using the Cox PH model in Chapter 5. Unlike the analysis in Chapter 5, the quantile analysis suggests that there is no significant AGE60 effect, nor a significant AGE60×SEX interaction. One possible interpretation is that these effects are only needed in the Cox PH model to adjust for a more general form of heterogeneity. Specifically, these effects might be most important for the larger responses, but might not be significant since the quantile analysis provides a defective distribution (*i.e.*, several of the largest survival times are censored). In fact, the largest τ probability is about .79. This indicates that there is little information in the data concerning the right tail of the distribution (say, $\tau > .75$). Thus, a regression quantile analysis might miss the AGE60 and AGE60×SEX effects that appear only in the higher quantiles, even though these would appear in a Cox analysis since the Cox PH model combines all the information in a single estimate. Therefore, the fact that the Cox approach finds AGE60 and AGE60×SEX effects could be an instance of the comment in Section 8.4 that Cox methods are more powerful when the model holds, but may miss important sources of heterogeneity. Note: when both approaches are used with only the three main effects (KPS.PRE., GROUP, and SEX), the two analyses are not very dissimilar. The KPS.PRE effect in the regression quantile model is significant for $\tau < .6$, though not as significant as indicated by the Cox analysis; the GROUP effect just fails to be significant at the 5% level for the 30th percentile and smaller, while it is significant for the Cox analysis; and the SEX effect (which had a p-value of .052 under the Cox PH model) is highly significant for smaller τ but not at all significant for larger τ. The details of the analysis from which these conclusions were drawn are provided in Section 8.6.2.

The second example comes from the text by Hosmer and Lemeshow (1999). It concerns the comparison of two courses of treatments for drug abuse ("long" and "short"). The dependent variable is the time to relapse. Here, the quantile analysis suggests that an important variable related to treatment compliance had been ignored. Inclusion of this covariate provides an example where the Cox PH and regression quantile models differ significantly. The regression quantile analysis provides much more significance for the treatment effect (once the significant "compliance-related" covariate is included). It also suggests that the population is heterogeneous in that the coefficient for the "compliance-related" covariate is not constant in τ, the effect being greater for those with shorter times to relapse ($\tau < .2$) and smaller for those with longer periods before relapse ($\tau > .7$). This is essentially the opposite of what a Cox PH model would suggest for this effect. The treatment effect appears to be constant in τ, and thus independent of the response level. However, as noted above, it is significantly larger than what would be suggested by the Cox approach. The detailed analysis appears in Section 8.6.3.

8.6.1 Software for censored regression quantiles

Software for computing censored regression quantiles and carrying out appropriate inference is under active development. At the present time, software consisting of R programs are available at http://www.stat.uiuc.edu/~portnoy/steve.html and have been tested using R and S-PLUS under PC and UNIX environments. The current implementations follow the above algorithm using the R function crq. This function requires a model formula just like that in coxph, but generally incorporating a transformation of the response. For example, to use "log"-times (that gives an accelerated failure rate model), execute > out <- crq(Surv(log(time),status) $\sim$ model), where model is given by a model formula.

This program outputs a list consisting primarily of the matrix out$sol that has a column for each τ-breakpoint and rows corresponding to the breakpoint value of τ (row 1), the coefficient for each of the variables in the model (that is, $\hat{\beta}_{-j}(\tau)$ is in row $j+1$), and the **central quantile function**, $Q_0(\tau) \equiv \bar{x}'\hat{\beta}(\tau)$ (in the last row). Some other optional input parameters may be specified, and additional output provides a list of the $\hat{\tau}_j$ for all censored observations. This information permits the censoring weights to be defined, and allows alternative (weighted) regression quantile algorithms to be used.

To provide a simple example of the form of the output, consider a very simple quantile regression with 5 observations:

```
> x <- 1:5
> y <- exp(1 + x + .2*rnorm(5))
> s <- c(1,1,0,0,1)
> ex <- crq(Surv(log(y),s) ~ x)
```

```
> ex$sol
          [,1]         [,2]         [,3]
[1,]  0.0000000  0.3333333  1.0000000
[2,]  0.4816187  0.8141252  0.8141252
[3,]  1.1322560  1.0657547  1.0657547
[4,]  3.5953226  3.7449506  3.7449506
```

The first row of ex$sol is the τ breakpoints. There is only one jump in the quantile function at $\tau = .3333333$. The quantile at $\tau = 0$ has intercept and slope approximately equal to .482 and 1.132, respectively. The value of the $(\tau = 0)$ regression line at the weighted mean of the x-values ($\bar{x} = 2.75$) is 3.595. This is the value of the central quantile function for $0 \leq \tau < .3333333$. This central quantile function corresponds to the inverse of the baseline survival function in a Cox PH model. The regression quantile line does not change until $\tau = .3333333$. The next quantile fit (for all $\tau \geq .3333333$) has intercept $= .814$ and slope $= 1.066$ with 3.745 being the value of this line at $\bar{x} = 2.75$ (to three decimal places). From this output, one could plot the data and regression quantile fits to see when the censored observations are crossed. It is easy here to see that the fourth observation lies below all the data, and so it was deleted in the algorithm. Only the third observation is actually crossed at $\tau = .333333$ and this crossing value can be used to define the censoring weights (as a function of τ) with $\hat{\tau}_1 = .333333$. See (8.16), Section 8.5.1.

The current software uses a bootstrap method for inference. For this purpose, the user must specify the number of bootstrap resamples (B) and a vector of probabilities $(\tau_1, \ldots, \tau_k)$. The routine described more completely in the examples below then resamples the data (with replacement) B times, calls crq, and saves the regression coefficients $\{\hat{\underline{\beta}}^*(\tau) : \tau = \tau_1, \cdots, , \tau_k\}$ for all of the bootstrap resamples. Using these resample-based regression coefficients, confidence intervals are produced by applying a variant of the bootstrap **Interquartile Range** method that permits confidence intervals to be asymmetric. For details, see Portnoy (2003). These confidence intervals seem to work well in practice so far, but there are a large number of possible alternatives that require investigation. Ongoing research may find some of these alternatives to be preferable. R functions are also available for producing tables of the confidence intervals and for producing plots of the $\hat{\underline{\beta}}(\tau)$ functions and of the central survival function, the inverse of the conditional quantile function at the mean vector $\bar{x}$. By default, these plots also graph results from fitting the Cox PH model for purposes of comparison.

8.6.2 CNS lymphoma example

The first example is the CNS lymphoma data for which Cox PH models were fit in Chapter 5. For simplicity, we will use the usual "log" transformation suggested by the accelerated failure rate model and fit $\log(T)$ in terms of the explanatory variables discussed in Chapter 5. Given only 58 observations, it is not surprising that both models tend to fit adequately and generally tell a somewhat similar story. However, as noted above, the ability of the regression quantile model to allow for much more general heterogeneity comes at a cost in statistical accuracy. Here, when we use model selected in Chapter 5 (explanatory variables: KPS.PRE., GROUP, SEX, AGE60, and SEX×AGE60), the regression quantile analysis suggested that the AGE60 and SEX×AGE60 were no longer statistically significant for any τ. It also generally showed less significance for the other coefficients. Otherwise, the analysis suggested that there was little difference between the Cox PH model and the censored regression quantile model; and in fact there appeared to be little heterogeneity in the data once the log(T) transformation was taken.

To present a more detailed example, we will therefore use the slightly simpler model suggested above based only on the explanatory variables: KPS.PRE., GROUP, and SEX. Specifically, the model we fit is:

$$\log(B3TODEATH) \sim \beta_0 + \beta_1\,KPS.PRE. + \beta_2\,GROUP + \beta_3\,SEX\,. \quad (8.18)$$

The first step is to use crq to compute the censored regression quantile coefficients $\hat{\underline{\beta}}_j(\tau)$.

```
> attach(cns2)
> cns2.crq <- crq(Surv(log(B3TODEATH),STATUS) ~ KPS.PRE.+GROUP
                    +SEX)
```

This produces the solution matrix cns2.crq$sol as described above. Here, cns2.crq$sol is a 6 by 61 matrix. The first row (the τ breakpoints) has largest value .7868, indicating that there are censored observations above all uncensored ones so that the survival distribution is defective. The next four rows correspond to the coefficients for the INTERCEPT, KPS.PRE., GROUP, and SEX, respectively; and the last row gives the central quantile.

Generally, we may view the regression quantile coefficients $\hat{\beta}_j(\tau)$ by plotting the values in the $(j+1)$th row of cns2.crq$sol against the first row (and connecting the points linearly). The central quantile function may be inverted to obtain a regression quantile version of a central survival function. This is an analog of the Cox PH survival function evaluated at $\underline{x} = \bar{\underline{x}}$, referred to as the Cox PH baseline survival function. The two may be compared. Specifically, the **central regression quantile survival function** is

$$\hat{S}(t) = 1 - \hat{Q}_0^{-1}(e^t), \quad (8.19)$$

where the exponentiation is needed to reverse the logarithm transformation in (8.18).

At this point, it would be useful to look at the coefficients $\hat{\underline{\beta}}_j(\tau)$, but we first need some measure of statistical accuracy. The bootstrap samples needed to compute confidence intervals as described above may be obtained using the function crq.boot. Unfortunately, the repeated samples in the bootstrap sometimes lead to serious degeneracy problems. One easy way to adjust for this is to "dither"; that is, add purely random noise that is smaller than the accuracy of the data. Thus, the following R commands may be run:

```
> Kd <- KPS.PRE. + .0001*rnorm(58)
> Gd <- GROUP + .0001*rnorm(58)
> Sd <- SEX + .0001*rnorm(58)
> dith <- crq(Surv(log(B3TODEATH),STATUS) ~ Kd+Gd+Sd)
> cns2.b <- crq.boot(dith,400,.1*(1:8))
```

This produces a bootstrap sample of size 400 of each of the coefficients in the model (the intercept and the 3 explanatory coefficients) at each of the deciles ($\tau = .1, .2, \cdots, .8$). A plotting function can now be used to graph $\hat{\beta}_j(\tau)$. The function that does this also produces an analog of the effect of the jth explanatory variable on the conditional quantile function as given by the Cox PH model. Basically, this is just an estimate of the derivative of the Cox conditional quantile (8.15) with respect to $x^{(j)}$ (the jth explanatory variable). Specifically, the vector of analogous Cox PH effects is

$$\underline{b}_{Cox}(\tau) \equiv \frac{\partial}{\partial \underline{x}} Q_{Cox}(\tau|\underline{x}) \mid_{\underline{x}=\bar{\underline{x}}} . \tag{8.20}$$

The following commands produce Figure 8.7, which gives plots for the regression quantile coefficients for each of the explanatory variables in model (8.18).

```
> cns2.cox <- coxph(Surv(B3TODEATH,STATUS) ~ KPS.PRE.+GROUP
                    +SEX)
> pltbeta.IQR(cns2.crq$sol,cns2.b,1,cns2.cox,"CNS2")
> pltbeta.IQR(cns2.crq$sol,cns2.b,2,cns2.cox,"CNS2")
> pltbeta.IQR(cns2.crq$sol,cns2.b,3,cns2.cox,"CNS2")
```

In each plot in Figure 8.7, the small dashed line is the conditional quantile coefficient $\hat{\beta}_j(\tau)$ as described above. The solid line is just a smoothed version of this (using the function supsmu). The large dashed line is the estimated Cox effect (8.20) for the corresponding variable. The shaded area represents the confidence bands for the regression quantile coefficients as described above. Note that since the Cox PH model imposes a fixed shape for the conditional quantile, this line must have the same form in each of the plots. Specifically, any pair of such "Cox-effect" curves are proportional; or equivalently, the ratio of the curves for any pair of variables must be a constant independent of τ.

What is most clear here is that the confidence bands are wider than any differences in the plots. The regression quantiles seem somewhat constant, and seem to have a slightly different shape from the analogous Cox effects. There may be a decline in the KPS.PRE. and, especially, the SEX coefficients: these coefficients seem to be somewhat larger for small τ and nearly zero for large τ. Such behavior could not have been captured using the Cox PH model. In general, the regression quantile effects appear to be larger in

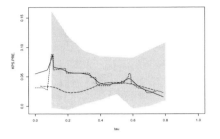

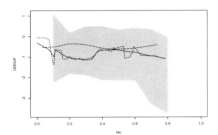

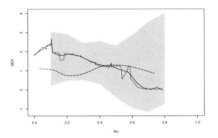

Figure 8.7 *Censored regression quantile functions for each of the coefficients in (8.18) for CNS data. The small dashed line is the raw estimates, the solid line smooths these estimates, and the large dashed line is the analogous Cox PH effect. Shading gives 95% pointwise confidence bands for the quantile functions as described above.*

absolute value than the analogous Cox effects though the difference is not statistically significant. Thus, there is some suggestion of heterogeneity in the model for log(B3TODEATH), and some hint that the Cox PH model may be missing something, but there is not sufficient statistical accuracy to make these features statistically significant.

The central regression quantile survival function (8.19) is compared with the Cox baseline survival function in Figure 8.8. Here the shaded region is the Cox PH model confidence bands, and the dotted lines are the upper and lower bands from the censored regression quantile analysis. Clearly there is little real difference. Lastly, we compare the individual coefficients numerically. The summary of the analysis of the Cox PH model is

```
> summary(cns2.cox)
          coef exp(coef) se(coef)     z       p
KPS.PRE. -0.0347    0.966    0.010 -3.45 0.00056
   GROUP  0.7785    2.178    0.354  2.20 0.02800
     SEX -0.7968    0.451    0.410 -1.94 0.05200
```

The Cox analysis provides strong statistical significance for KPS.PRE. and mild significance for GROUP, with SEX just missing significance at the .05

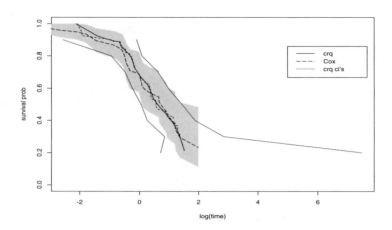

Figure 8.8 *Central regression quantile survival function for CNS data as computed by* crq *with the Cox PH baseline survival function. The shaded region gives the 95% confidence bands for the Cox PH estimate.*

level. For regression quantiles, we must generally look at coefficients for multiple values of τ. We list results for the 30th, 50th (median), and 70th-percentiles (that is, corresponding to the third, fifth, and seventh indices of the τ-vector specified in the crq.boot computation):

```
> sumpr.crq(ci.boot(cns2.crq$sol,cns2.b,c(3,5,7)))
 tau = 0.3     95% CI
            COEF     LOWER     UPPER     Z-STAT    PROB
     INT -4.86428  -8.38705  -0.47820  -2.41100  0.01591
 KPS.PRE.  0.05616  0.00369   0.09410   2.43480  0.01490
   GROUP -1.15368  -1.89674   0.49020  -1.89470  0.05814
     SEX  1.73653   0.51390   2.44730   3.52080  0.00043

 tau = 0.5     95% CI
            COEF     LOWER     UPPER     Z-STAT    PROB
     INT -2.80811  -6.56808   0.11712  -1.64659  0.09964
 KPS.PRE.  0.03855   0.02051   0.08354   2.39748  0.01651
   GROUP -0.59175  -2.13677   0.31737  -0.94520  0.34455
     SEX  1.24505  -0.38702   2.31650   1.80528  0.07103

 tau = 0.7     95% CI
            COEF     LOWER     UPPER     Z-STAT    PROB
     INT -0.14279  -6.793e+00  1.86465  -0.06465  0.94845
 KPS.PRE.  0.02245  -2.755e-05  0.09584   0.91783  0.35871
   GROUP -0.96943  -3.322e+00  0.33687  -1.03866  0.29896
     SEX  0.04948  -1.161e+00  3.56127   0.04108  0.96724
```

For the 30th-percentile responses, the regression quantile analysis shows the GROUP and KPS.PRE. effects to be significant at very similar levels to those given by the Cox analysis. However, the SEX effect is much more significant in this lower percentile than is indicated by the Cox analysis. So for shorter survival times (specifically, the 30th-percentile), on the average women live $e^{1.74} = 5.7$ times longer than men with other factors fixed. The Cox PH model suggests that women have about .45 of the hazard rate of men with other factors fixed, but doesn't apply directly to survival times. The SEX effect in the Cox PH model is also weaker, just missing significance at the .05 level.

The median ($\tau = .5$) regression quantile still shows a significant KPS.PRE. effect, but the GROUP effect is no longer significant, and the SEX effect just missed significance at the 5% level. Perhaps surprisingly, none of the x-variables are significant for the 70th-percentile. This is because there is little information at the larger percentiles – recall that the distribution is actually defective above .79 since so many of the largest observations are censored. As noted in the introduction to this section, since the Cox PH model is global, it does not need information in this right tail; but for the same reason, it is unable to find the greater significance in the smaller (and middle) percentiles.

8.6.3 UMARU impact study (UIS) example

The following example is taken from the text by Hosmer and Lemeshow (1999). It concerns the comparison of two courses of treatments for drug abuse ("long" = 1 and "short" = 0). The dependent variable is the time to relapse, with those not relapsing before the end of the study considered as censored. This data set contains complete records for 575 subjects and 8 independent variables. In addition to the treatment effect, TREAT ("1" for 6 month treatment, "0" for 3 month treatment), the independent variables were the following covariates: AGE (in years), BECK (a depression score at admission to the program), IV3 (1 if there was recent IV drug use, 0 otherwise), NDT (the number of prior drug treatments), RACE ("white" or "other"), SITE ("A" or "B"), and LOT (number of days on treatment). The text develops a model with NDT transformed into a two-dimensional variable (ND1 and ND2), and did not use the "compliance" indicator, LOT. Their "best model" used TREAT, AGE, BECK, IV3, ND1, ND2, RACE, SITE, AGE × SITE, and RACE × SITE.

Use of the Cox PH model is described extensively in Hosmer and Lemeshow. Here a censored regression quantile analysis is performed on log(TIME). Using the design matrix for the text's model, the regression quantile analysis showed almost exactly the same results as the Cox PH model. The quantile function estimates and their confidence bands were barely distinguishable when plotted, and there were no significant differences between $\hat{\beta}(\tau)$ and the Cox $\underline{\beta}$-effect estimates as defined by (8.20).

However, note that the text did not use the "compliance" indicator: the number of days actually on treatment. Indeed, introduction of such a variable generally complicates the interpretation of the treatment effect. Nonetheless, in an effort to use this information, consider using the fraction of the assigned treatment time actually administered. That is, define FRAC to be LOT/90 for the "short" treatment and LOT/180 for the "long" treatment. It turned out that FRAC was only mildly correlated with the treatment effect, thus suggesting that the treatment effect can be interpreted independently of FRAC. Introduction of the FRAC variable provides models that are clearly better, both for regression quantiles and for the Cox PH model: the test statistics for testing the entire model nearly triple. When FRAC is included in the model and the RACE × SITE interaction is deleted (since it is now highly insignificant), a rather different comparison emerges. Figure 8.9 presents the central

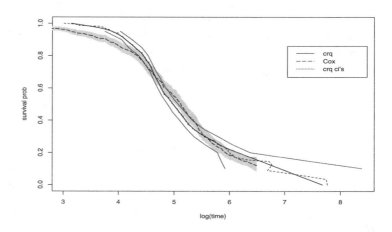

Figure 8.9 *The central survival function for the Drug relapse data from Hosmer and Lemeshow (1999) as computed by* crq *with a bootstrap 95% pointwise confidence band indicated by the dotted line. The shaded region gives the 95% confidence bands for the Cox PH estimate.*

and baseline survival function estimates. Although the censored quantile and Cox estimates tell a similar qualitative story, it is clear that two estimates differ, especially in the left tail. The confidence interval widths are somewhat similar, but the Cox baseline estimate has a much heavier left tail. These differences would be quite clear in density or hazard function estimates.

Figure 8.10 summarizes the quantile coefficient effects: $\hat{\beta}(\tau)$ for regression quantiles and $\underline{b}(\tau)$ from (8.20) for the Cox approach. The $\overline{\text{Cox}}$ estimate of the treatment coefficient is $\hat{\beta}_1 = .49$ (std. err. $= .10$), indicating that the hazard rate to remain drug-free is $e^{.49} = 1.6$ times greater for patients on the long treatment than for those on the short treatment (on the average, other factors

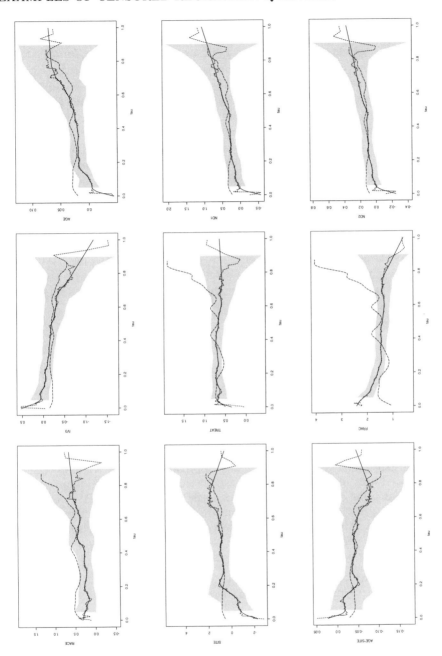

Figure 8.10 *Quantile regression coefficients for the Drug relapse data from Hosmer and Lemeshow (1999), with shaded 95% confidence bands. The large dashed line is the Cox effect (8.20).*

kept fixed). The treatment quantile effect is in the fifth panel: the Cox PH quantile effect (8.20) is much smaller for $\tau < .8$. The regression quantile approach suggests a definitely stronger treatment effect: in fact, for $\tau < .6$, the z-statistic for $\hat{\beta}(\tau)$ is roughly twice as large as the Cox z statistic ($z = 10.8$ for the regression quantile at $\tau = .4$ *vs.* $z = 5.1$ for the Cox effect). The regression quantile coefficient is quite constant for $\tau < .8$, and is generally greater than .65. Since the response is "log(survival time)," this indicates that patients on the long treatment remain drug-free nearly twice as long as those on the short treatment (other factors kept fixed). The effect of the fraction of treatment completed (FRAC) also shows substantial differences between the approaches. The regression quantile model indicates that the response to FRAC decreases as τ increases (that is, it tends to be larger for those with smaller relapse times), and that it is much larger than suggested by the Cox PH model (except perhaps for very large τ). The greater flexibility provided by regression quantile appears to permit a more sensitive analysis of the treatment effect. Specifically, the quantile restrictions imposed by the Cox PH model appear to confound the treatment and FRAC effects, making the treatment effects appear less significant.

8.7 Exercises

A. *Applications*

8.1 For the sample, $y_i = 1, 2, 3, 4, 5$, plot $\sum_{i=1}^{5} \rho_\tau(y_i - \xi)$, the objective function (8.3), as a function of ξ for $\tau = .3$ and $\tau = .5$. Plot the objective function for random samples rnorm(20) and runif(100) for $\tau = .3$ and $\tau = .5$.

8.2 Review the ADDICTS data of Chapter 7.1. Although the earlier analysis suggested an extended Cox model approach, here we compare the Cox PH analysis of Part II with a censored regression quantile approach. The output using censored regression quantile analysis follows. We also replace "dose" with "log(dose)" and present the output using both coxph and crq, crq.boot. Interpret the results and compare the two approaches. Is use of "log(dose)" preferable to "dose" for the regression quantile approach and/or for the Cox approach?

```
> attach(add.d)
> add.crq <- crq(Surv(log(time),status) ~ clinic+prison
                 +dose)
> cd <- clinic + .0001*rnorm(238)
> pd <- prison + .0001*rnorm(238)
> dd <- dose + .0001*rnorm(238)
> dith <- crq(Surv(log(time),status) ~ cd + pd + dd)
> add.b <- crq.boot(dith, 400, .05*(1:13))
> sumpr.crq(ci.boot(add.crq$sol,add.b,c(5,8,10,12)))
```

```
tau = 0.25    95% ci
            COEF    LOWER    UPPER    Z-STAT    PROB
   INT    3.42390  1.65476  4.90419  4.13048  1.8100e-05
   clinic 0.29668 -0.11256  0.78374  1.29754  9.7222e-02
   prison -0.48268 -0.81721 -0.03592 -2.42175  7.7230e-03
   dose   0.03195  0.01751  0.05802  3.09122  9.9667e-04


tau = 0.4    95% ci
            COEF    LOWER    UPPER    Z-STAT    PROB
   INT    3.37272  2.32915  4.26161  6.84156  3.9167e-12
   clinic 0.57125  0.04284  1.01549  2.30228  1.0660e-02
   prison -0.22725 -0.80909  0.26516 -0.82926  2.0348e-01
   dose   0.03232  0.02306  0.04951  4.78956  8.3573e-07


tau = 0.5    95% ci
            COEF    LOWER    UPPER    Z-STAT    PROB
   INT    3.52229  2.89783  4.51857  8.51920  0.0000e+00
   clinic 0.71024 -0.01413  1.40258  1.96520  2.4694e-02
   prison -0.23882 -0.47497  0.12842 -1.55150  6.0386e-02
   dose   0.03098  0.02299  0.03795  8.11770  2.2204e-16


tau = 0.6    95% ci
            COEF    LOWER    UPPER    Z-STAT    PROB
   INT    4.06809  2.39009  4.92696  6.28610  1.6281e-10
   clinic 0.74531  0.20936  1.32380  2.62160  4.3759e-03
   prison -0.20959 -0.43355  0.12202 -1.47880  6.9591e-02
   dose   0.02371  0.01686  0.03958  4.09120  2.1455e-05


> add.cox4 <- coxph(Surv(time,status) ~ clinic+prison
                    +log(dose))
> add.crq4 <- crq(Surv(log(time),status) ~ clinic+prison
                  +log(dose),mw=100)
> add.b4 <- crq.boot(add.crq4,400,.05*(1:13),mw=100)
> summary(coxph(Surv(time,status)~clinic+prison+log(dose)))


            coef  exp(coef)  se(coef)    z      p
   clinic -1.043    0.352     0.215   -4.85  1.3e-06
   prison  0.348    1.416     0.168    2.07  3.8e-02
log(dose) -1.867    0.155     0.331   -5.65  1.6e-08


         exp(coef)  exp(-coef)  lower .95   upper .95
   clinic   0.352     2.837      0.2311      0.537
   prison   1.416     0.706      1.0188      1.967
log(dose)   0.155     6.471      0.0808      0.295


Rsquare= 0.232 (max possible= 0.997)
Likelihood ratio test= 62.8 on 3 df, p=1.52e-13
Wald test            = 56.1 on 3 df, p=3.91e-12
Efficient score test = 58.1 on 3 df, p=1.5e-12
```

```
> print(cox.zph(add.cox4))
                 rho   chisq        p
   clinic   -0.2626  11.689  0.000629
   prison   -0.0437   0.287  0.592354
log(dose)    0.0562   0.376  0.539872
   GLOBAL        NA  12.999  0.004638


> sumpr.crq(ci.boot(add.crq4$sol,add.b4,c(5,8,10,12)))


tau = 0.25       95% ci
                COEF     LOWER     UPPER    Z-STAT      PROB
       INT  -2.88006  -8.89829   1.08524   -1.1308   0.25812
    clinic   0.29180   0.01799   0.94047    1.2399   0.21499
    prison  -0.49407  -0.86352  -0.16336   -2.7662   0.00567
 log(dose)   2.01427   1.14010   3.50626    3.3370   0.00085


tau = 0.4        95% ci
                COEF     LOWER     UPPER    Z-STAT      PROB
       INT  -2.37891  -9.46431   1.24350   -0.8709   0.38382
    clinic   0.50623   0.00287   1.24570    1.5967   0.11034
    prison  -0.26376  -0.89769   0.13690   -0.9994   0.31760
 log(dose)   1.90928   0.98582   3.38150    3.1242   0.00178


tau = 0.5        95% ci
                COEF     LOWER     UPPER    Z-STAT      PROB
       INT  -0.12469  -4.63002   1.91981   -0.0746   0.94051
    clinic   0.63428  -0.10334   1.21299    1.8889   0.05891
    prison  -0.20672  -0.53684   0.10206   -1.2683   0.20468
 log(dose)   1.37392   0.86498   2.44177    3.4156   0.00064


tau = 0.6        95% ci
                COEF     LOWER     UPPER    Z-STAT        PROB
       INT   0.15301  -2.04626   1.15444    0.1874  8.5135e-01
    clinic   0.74609   0.24016   1.07773    3.4919  4.7965e-04
    prison  -0.23769  -0.47199  -0.03798   -2.1468  3.1806e-02
 log(dose)   1.31160   1.07247   1.74661    7.6267  2.3981e-14
```

B. *Theory and WHY!*

8.3 Prove that expression (8.8) gives the quantile objective function at a solution.

8.4 For a single sample $Y_1, \ldots, Y_n$ with order statistics $Y_{(1)} < Y_{(2)} < \ldots < Y_{(n)}$, show that the ith largest observation, $Y_{(i)}$, is the τth-quantile if and only if $(i-1)/n \le \tau \le i/n$. Use the subgradient inequality: specifically, show that the left derivative (in the negative direction) is nonnegative if and only if $-(i-1)(1-\tau) + (n-i+1)\tau \ge 0$, and the right derivative (in the positive direction) is nonnegative if and only if $i(1-\tau) - (n-i)\tau \ge 0$.

8.5 Consider the simple regression through the origin. Take a sample (x_i, y_i) and note that each observation determines a line through the origin (and through that point) with slope $b_i = y_i/x_i$. You may assume that the $\{x_i\}$ are all different positive values. Let $\{b_{(i)}\}$ denote the ordered values of the slopes (in increasing order), and let x_i^* denote the x_i value corresponding to $b_{(i)}$ (note: x_i^* is not the ordered value of x_i, but is the denominator of $b_{(i)}$). Let $s = \sum_{i=1}^{n} x_i$. Let $\hat{\beta}(\tau)$ be the regression quantile coefficient; that is, the value of b achieving $\min_b \sum_{i=1}^{n} \rho_\tau(y_i - bx_i)$.

(a) Show that $\hat{\beta}(\tau) = b_{(i)}$ if and only if the following two subgradient inequalities hold:

$$-(1 - \tau) \sum_{j=1}^{i-1} x_j^* + \tau \sum_{j=i}^{n} x_j^* \;\geq\; 0$$

$$(1 - \tau) \sum_{j=1}^{i} x_j^* - \tau \sum_{j=i+1}^{n} x_j^* \;\geq\; 0 \;.$$

As a consequence, show that $\hat{\beta}(\tau) = b_{(i)}$ if and only if

$$\frac{\sum_{j=1}^{i-1} x_j^*}{s} \leq \tau \leq \frac{\sum_{j=1}^{i} x_j^*}{s} \;.$$

(b) For the sample of four pairs: $\{(x_i, y_i) = (1, 2), (2, 3), (3, 9), (4, 5)\}$, compute $\hat{\beta}(\tau)$ for all τ. Plot the unique regression quantile lines and indicate the corresponding τ-intervals.

Hint: make a table of the ordered slope values, the corresponding x_i^*, and the partial sums of the $\{x_i^*\}$.

(c) Do the same as in (b) above for $\{(x_i, y_i) = (-1, 1), (1, 1), (2, 3), (3, 2)\}$.

Hint: Is the first observation (-1, 1) a solution for any τ?

8.6 Consider the general regression quantile setup, but extend the notation so that $\hat{\beta}$ depends explicitly on the data: that is, write $\hat{\beta}(\tau; X, \underline{Y})$ for $\hat{\beta}(\tau)$, where X is the design matrix with rows $\underline{x}_i$, and the responses are in the vector $\underline{Y}$.

(a) Using the objective function (8.6), show the following:

$$\begin{aligned}
\underline{\beta}(\tau; X, c\underline{Y}) &= c\hat{\beta}(\tau; X, \underline{Y}) & \text{for } c > 0 \\
\underline{\beta}(\tau; X, \underline{Y} + X\underline{a}) &= \underline{\beta}(\tau; X, \underline{Y}) + \underline{a} & \text{for } \underline{a} \in \Re^m \;.
\end{aligned}$$

(b) Show that the subgradient inequalities depend on the y_i values **only** through the sign of the residuals. As a consequence, if y_i is perturbed so that $(y_i - \underline{x}_i'\underline{\beta}(\tau))$ keeps the same sign, then $\underline{\beta}(\tau)$ remains unchanged.

References

Aalen, O.O. (1978). Nonparametric inference for a family of counting processes. *Ann. Statist.*, **6**, 534 – 545.

Agresti, A. (1990). *Categorical Data Analysis*. New York: Wiley-Interscience.

Akaike, H. (1974). A New Look at the Statistical Model Identification. *IEEE Trans. Automat. Contr.*, **AC-19**, 716 – 723.

Andersen, P.K and Gill, R.R. (1982). Cox's regression model for counting processes: A large sample study. *Ann. Statist.*, **10**, 1100 – 1120.

Babu, G.J. and Feigelson, E.D. (1996). *Astrostatistics*, London: Chapman & Hall.

Belsley, D.A., Kuh, E., and Welsch, R.E. (1980). *Regression Diagnostics: Identifying Influential Data and Sources of Collinearity*. New York: Wiley.

Bickel, P.J. and Doksum, K.A. (2001). *Mathematical Statistics: Basic Ideas and Selected Topics, Vol.I, 2nd Edition*. New Jersey: Prentice-Hall, Inc.

Breslow, N. and Crowley, J. (1974). A large sample study of the life table and product limit estimates under random censorship. *Ann. Statist.*, **2**, 437 – 453.

Cain, K.C. and Lange, N.T. (1984). Approximate case influence for the proportional hazards regression model with censored data. *Biometrics*, **40**, 493 – 499.

Caplehorn, J., et al. (1991). Methadone dosage and retention of patients in maintenance treatment. *Med. J. Australia*, **154**, 195 – 199.

Collett, D. (1999). *Modelling Survival Data in Medical Research*. London: Chapman & Hall/CRC.

Cook, R.D. and Weisberg, S. (1982). *Residuals and Influence in Regression*. London: Chapman & Hall.

Cox, D.R. (1959). The analysis of exponentially distributed life-times with two types of failure. *J.R. Statist. Soc.*, **B, 21**, 411 – 421.

Cox, D.R. (1972). Regression models and life-tables (with discussion). *J.R. Statist. Soc.*, **B, 34**, 187 – 220.

Cox, D.R. (1975). Partial likelihood. *Biometrika*, **62**, 269 – 276.

Cox, D.R. and Oakes, D. (1984). *Analysis of Survival Data*. London: Chapman & Hall.

Cox, D.R. and Snell, E.J. (1968). A general definition of residuals (with discussion). *J.R. Statist. Soc.*, **A, 30**, 248 – 275.

Dahlborg, S.A., Henner, W. D., Crossen, J.R., Tableman, M., Petrillo, A., Braziel, R. and Neuwelt, E.A. (1996). Non-AIDS primary CNS lymphoma: the first example of a durable response in a primary brain tumor using enhanced chemotherapy delivery without cognitive loss and without radiotherapy. *The Cancer Journal from Scientific American*, **2**, 166 – 174.

Davison, A.C. and Hinkley, D.V. (1999). *Bootstrap Methods and their Application*. London: Cambridge University Press.

DeGroot, M.H. (1986). *Probability and Statistics, 2nd Edition.* New York: Addison-Wesley.

Edmunson, J.H., Fleming, T.R., Decker, D.G., Malkasian, G.D., Jefferies, J.A., Webb, M.J., and Kvols, L.K. (1979). Different chemotherapeutic sensitivities and host factors affecting prognosis in advanced ovarian carcinoma vs. minimal residual disease. *Cancer Treatment Reports,* **63**, 241–47.

Efron, B. (1967). The two sample problem with censored data. *Proc. Fifth Berkeley Symposium in Mathematical Statistics,* **IV**, New York: Prentice-Hall, 831 – 853.

Efron, B. (1979). Bootstrap methods: another look at the jackknife. *Ann. Statist.,* **7**, 1 – 26.

Efron, B. (1998). R. A. Fisher in the 21st Century. *Statist. Sci.,* **13**, 95 – 122.

Efron, B. and Gong, G. (1983). A leisurely look at the bootstrap, the jackknife, and cross-validation. *Amer. Statist.,* **37**, 36 – 48.

Efron, B. and Petrosian, V. (1992). A simple test of independence for truncated data with applications to red shift surveys, *Astrophys. J.,* **399**, 345 – 352.

Efron, B. and Tibshirani (1993). *An Introduction to the Bootstrap.* London: Chapman & Hall.

Embury, S.H., Elias, L., Heller, P.H., Hood, C.E., Greenberg, P.L., and Schrier, S.L. (1977). Remission maintenance therapy in acute myelogenous leukemia. *Western Journal of Medicine,* **126**, 267 – 272.

Finkelstein, D.M., Moore, D.F., and Schoenfeld, D.A. (1993). A proportional hazards model for truncated AIDS data. *Biometrics,* **49**, 731 – 740.

Fleming, T. and Harrington, D. (1991). *Counting Processes and Survival Analysis.* New York: Wiley.

Galton, F. (1889). *Natural Inheritance.* London: Macmillan.

Gehan, E.A. (1965). A generalized Wilcoxon test for comparing arbitrarily singly-censored samples. *Biometrika,* **52**, 203 – 223.

Gooley, T.A., Leisenring, W., Crowley, J., and Storer, B.E. (1999). Estimation of failure probabilities in the presence of competing risks: new representations of old estimators. *Statist. Med.,* **18**, 695 – 706.

Gooley, T.A., Leisenring, W., Crowley, J.C., and Storer, B.E. (2000). Why the Kaplan-Meier fails and the cumulative incidence function succeeds when estimating failure probabilities in the presence of competing risks. Editor: J.C. Crowley. *Handbook of Statistics in Clinical Oncology.* New York: Marcel Dekker, Inc., 513 – 523.

Grambsch, P. and Therneau, T.M. (1994). Proportional hazards tests and diagnostics based on weighted residuals. *Biometrika,* **81**, 515 – 526.

Gray, R.J. (2002). cmprsk library, competing risks library for S-PLUS. http://biowww.dfci.harvard.edu/~gray/.

Gray, R.J. (2002). cmprsk.zip, competing risks R library. http://www.r-project.org/~CRAN/.

Greenwood, M. (1926). The natural duration of cancer. *Reports on Public Health and Medical Subjects,* **33**, 1 – 26, London: Her Majesty's Stationery Office.

Gutenbrunner, C., Jurečková, J., Koenker, R., and Portnoy, S. (1993). Tests of linear hypotheses based on regression rank scores. *J. Nonparametric Statist.,* **2**, 307–331.

Hoel, D.G. and Walburg, H.E., Jr. (1972). Statistical analysis of survival experiments. *J. Natl. Cancer Inst.,* **49**, 361 – 372.

Hogg, R.V. and Craig, A.T. (1995). *Introduction to Mathematical Statistics, 5th Edition.* New Jersey: Prentice Hall.

Hosmer, D.W. Jr. and Lemeshow, S. (1999). *Applied Survival Analysis: Regression Modeling of Time to Event Data.* New York: Wiley.

Kaplan, E.L. and Meier, P. (1958). Nonparametric estimation from incomplete observations. *J. Amer. Statist. Assoc.*, **53**, 457 − 481.

Kalbfleisch, J.D. and Prentice, R.L. (1980). *The Statistical Analysis of Failure Time Data.* New York: Wiley.

Keiding, N. (1992). Independent delayed entry, in *Survival Analysis: State of the Art*, J.P. Klein and P. Goel, eds. Boston: Kluwer Academic Publishers, 309 − 326.

Klein, J.P. and Moeschberger, M.L. (1997). *Survival Analysis: Techniques for Censored and Truncated Data.* New York: Springer.

Kleinbaum, D.G. (1995). *Survival Analysis: A Self-Learning Text.* New York: Springer.

Koenker, R. (1994). Confidence intervals for regression quantiles, in *Asymptotic Statistics: Proc. 5th Prague Symposium*, editors: P. Mandl and M. Hušková. Heidelberg: Physica-Verlag.

Koenker, R. and Bassett, G. (1978). Regression quantiles. *Econometrica*, **46**, 33−50.

Koenker, R. and d'Orey, V. (1987). Computing regression quantiles. *Appl. Statist.*, **36**, 383 − 393.

Koenker, R., and Geling, O. (2001). Reappraising Medfly longevity: a quantile regression survival analysis. *J. Amer. Statist. Assoc.*, **96**, 458 − 468.

Koenker, R. and Machado, J. (1999). Goodness of fit and related inference procedures for quantile regression. *J. Amer. Statist. Assoc.*, **94**, 1296 − 1310.

Lawless, J.F. (1982). *Statistical Models and Methods for Lifetime Data.* New York: Wiley.

Lee, E.T. (1992). *Statistical Methods for Survival Data Analysis, 2nd Edition.* New York: John Wiley & Sons.

Leiderman, P.H., Babu, D., Kagia, J., Kraemer, H.C., and Leiderman, G.F. (1973). African infant precocity and some social influences during the first year. *Nature*, **242**, 247 − 249.

Lenneborg, C.E. (2000). *Data Analysis by Resampling: Concepts and Applications.* Pacific Grove: Duxbury.

MathSoft (1999). *S-PLUS 2000-Guide to Statistics.* Seattle, WA: MathSoft, Inc.

Mantel, N. and Haenszel, W. (1959). Statistical aspects of the analysis of data from retrospective studies of disease. *J. National Cancer Institute*, **22**, 719 − 322.

Miller, R.G. (1981). *Survival Analysis.* New York: Wiley.

Morrell, C.H. (1999). Simpson's Paradox: an example from a longitudinal study in South Africa. *J. Statist. Ed.*, **7**, **3**.

Nelson, W.B. (1972). Theory and applications of hazard plotting for censored failure data. *Technometrics*, **14**, 945 − 965.

Nelson, W.B. and Hahn, G.B. (1972). Linear estimation of regression relationships from censored data, part 1−simple methods and their applications (with discussion). *Technometrics*, **14**, 247 − 276.

Peterson, A.V. (1975). *Nonparametric Estimation in the Competing Risks Problem.* Ph.D.thesis, Department of Statistics, Stanford University.

Peto, R. (1973). Empirical survival curves for interval censored data. *Appl. Statist.*, **22**, 86 − 91.

Peto, R. and Peto, J. (1972). Asymptotically efficient rank invariant test procedures. *J.R. Statist. Soc.*, **A**, **135**, 185 − 198.

Pike, M.C. (1966). A method of analysis of certain class of experiments in carcinogenesis. *Biometrics*, **22**, 142 − 161.

Portnoy, S. (1991a). Asymptotic behavior of the number of regression quantile breakpoints. *SIAM J. Sci. Stat. Comp.*, **12**, 867 − 883.

Portnoy, S. (1991b). Behavior of regression quantiles in non-stationary, dependent cases. *J. Multivar. Anal.*, **38**, 100 − 113.

Portnoy, S. (2003). Censored regression quantiles. *J. Amer. Statist. Assoc.*, to appear.

Portnoy, S. and Koenker, R. (1997). The Gaussian Hare and the Laplacian Tortoise: computability of squared-error vs. absolute-error estimators. *Statist. Sci.*, **12**, 279 − 300.

Prentice, R.L. and Marek, P. (1979). A qualitative discrepancy between censored data rank tests. *Biometrics*, **35**, 861 − 867.

Reid, N. (1994). A conversation with Sir David Cox, *Statist. Sci.*, **9**, 439 − 455.

Ross, S.M. (2000). *Introduction to Probability Models, 7th Edition*. Orlando: Academic Press, Inc.

Schoenfeld, D.A. (1982). Partial residuals for the proportional hazards regression model. *Biometrika*, **69**, 239 − 241.

Simpson, E.H. (1951). The interpretation of interaction in contingency tables. *J.R. Statist. Soc.*, **B**, **13**, 238 − 241.

Smith, P.J. (2002). *Analysis of Failure and Survival Data*. Boca Raton: Chapman & Hall/CRC.

Therneau, T.M., Grambsch, P.M., and Fleming, T.R. (1990). Martingale-based residuals for survival models. *Biometrika*, **69**, 239 − 241.

Tsai, W.Y., Jewell, N.P., and Wang, M.C. (1987). A note on the product-limit estimator under right censoring and left truncation. *Biometrika*, **74**, 883 − 886.

Tsai, W.Y. (1990). The assumption of independence of truncation time and failure Time. *Biometrika*, **77**, 169 − 177.

Tsiatis, A. (1975). A nonidentifiability aspect of the problem of competing risks. *Proc. Natl. Acad. Sci.*, **72**, 20 − 22.

Tsuang, M.T. and Woolson, R.F. (1977). Mortality in patients with schizophrenia, mania and depression. *British Journal of Psychiatry*, **130**, 162 − 166.

van der Vaart, A.W. (1998). *Asymptotic Statistics*. London: Cambridge University Press.

Venables, W.N. and Ripley, B.D. (2002). *Modern Applied Statistics with S, 4th Edition*. New York: Springer-Verlag, Inc.

Woolson, R.F. (1981). Rank tests and a one-sample log rank test for comparing observed survival data to a standard population. *Biometrics*, **37**, 687 − 696.

Index